Mara Gabriela Novy Quadri
John Alexander Rodriguez Bustos

Use of natural extracts of cassava (Manihot esculenta Crantz)

Mara Gabriela Novy Quadri
John Alexander Rodriguez Bustos

Use of natural extracts of cassava (Manihot esculenta Crantz)

For treating water containing organic waste

ScienciaScripts

Imprint

Cover image: www.ingimage.com

This book is a translation from the original published under ISBN 978-3-330-76697-6.

Publisher:
Sciencia Scripts
is a trademark of
Dodo Books Indian Ocean Ltd. and OmniScriptum S.R.L publishing group

120 High Road, East Finchley, London, N2 9ED, United Kingdom
Str. Armeneasca 28/1, office 1, Chisinau MD-2012, Republic of Moldova, Europe
Managing Directors: Ieva Konstantinova, Victoria Ursu
info@omniscriptum.com

Printed at: see last page
ISBN: 978-620-8-62324-1

SUMMARY

To my parents Blanca and Luis, my brother Christian and my wife Paola, for their love, trust and unconditional support.

THANKS

To God, for another opportunity for professional and personal enrichment.

To my father Luis and my mother Blanca, the true authors of this work.

To my brother Christian for his support during the most difficult times and for his unconditional friendship.

To Pao for his love, dedication and affection and for believing in this adventure and his determination to always move forward together.

To the National Council for Scientific and Technological Development (CNPq) for the financial support through the Postgraduate Student Agreement Program (PEC-PG).

To my advisor, Prof. Dr. Mara Gabriela Novy Quadri. Dr. Mara Gabriela Novy Quadri for the knowledge she imparted and for her trust and credibility in my work.

To Prof. Regina de Fâtima P.M. Moreira for her support and help in the experimental development of the work and for her contributions.

To Prof. Dr. Antônio Augusto Ulson de Souza for his contributions and suggestions.

Edson Wuerges from EPAGRI - Santa Catarina for his collaboration in obtaining raw materials for the work

To Professor Joao Borges Laurindo and Engineer Giustino Tribuzi of the PROFI laboratory for their collaboration in the experimental part of this work.

To Engineer Alvaro Vargas Júnior of the Federal Institute of Santa Catarina (IFC) for his help in making samples.

To the LASIPO staff, especially Solange, Raquel, Marcel and Fabricio for their friendship and help at all times.

To the scientific initiation student Maria Julia das Chagas for her help in the experimental development of the work.

To the friends, family and colleagues who understood my absence and always tried to help me.

To the Colombian people in Florianópolis, especially Lore, Rafael, Julian and Leonardo for their friendship and companionship.

To Uncle Carlos, who left before his time and looks after me from heaven.

To my land Colombia for training good people To Brazil for welcoming me with affection.

SUMMARY

The aerial part of cassava *(Manihot Esculenta* Crantz*)*, a plant grown in Brazil, is little used and often discarded as waste. Its composition includes Iectins and polyphenols, compounds whose chemical structure is related to agglomeration phenomena and can therefore be used in various technological fields. In this study, aqueous and ethanolic extracts of cassava leaves and stems were applied to the treatment of effluents on a laboratory scale. The extracts were initially evaluated for protein and phenolic content, with the highest protein concentrations being obtained in the leaves (0.44 mg· $^{mL(-1)}$ of extract). The phenolic content was also determined, showing a high percentage of condensed tannins (90-100% of total tannins). With regard to the extracting agents (water and 70% ethanol), ethanol showed the best yield in both the leaves (59%) and the stem (37%), although its use was inconvenient in the case of the leaves, due to the extraction of undesirable pigments. The hemagglutinating activity of the extracts on various types of erythrocytes was evaluated, detecting agglutination in some of them (human, bovine and chicken) with different intensities, which suggests the presence of a specific lectin. Using significance tests and analysis of variance, the extraction (25°C and 60 min) and storage (10°C) conditions under which the agglutination phenomenon is greatest were established. Subsequently, the extracts were tested in the decolorization of aqueous solutions of methylene blue, finding a greater removal of color when the aqueous phase from the concentration of the ethanolic extract of the cassava stem was used. Once the extract had been chosen, coagulation tests were carried out to remove the dye and, using a fractional factorial design, different variables were evaluated in order to achieve maximum removal (0.10 mg dye· ml extract$^{-1)}$. The pH and extract dosage were the most influential factors. Subsequently, two other cationic dyes (Maxilon Blue GRL and Malachite Green) were tested, achieving maximum removals of 0.19 and 0.14 mg dye· ml extract^{-1} respectively. Neutral and basic pHs, low dye concentrations, extract dosages of around 40 ml· $^{L(-1)}$, absence of NaCl and stirring times of 5 and 30 min for the fast and slow mixing stages favored the decolorization process. In addition, the addition of an alkalizing agent was tested, which increased the removal rate to 90%. Finally, the chemical oxygen demand and toxicity of the solutions treated with the extract were evaluated. The COD increased and the toxicity of the treated samples decreased by 80%. The lethal concentration of the extract in the aqueous phase was 80 ml· $^{L\wedge(1)}$. As the dosages measured in the decolorization tests were lower than the lethal dosage determined, they were considered innocuous.

Keywords: Cassava. Lectins. Hemagglutinating activity. Tannins. Discoloration.

1 INTRODUCTION

The use of waste from agro-industrial production and the treatment of effluents are part of the growing global environmental concern; the constant search for sustainable processes is a necessity for today's industry.

Cassava *(Manihot Esculenta* Crantz*), a* plant native to the Brazilian Amazon [1], has an aerial part made up of leaves, stems and stalks, which is generally considered to be waste [2]. This part of the plant contains anti-nutritional components that limit its use as a food supplement [3]. Among these substances are lectins, glycoproteins with the ability to bind complex structures that have a sufficient number of free sites [4]. In addition, there are tannins, secondary phenolic components that play various roles in plant metabolism [5]. These two active ingredients have been linked to applications that include particle agglomeration phenomena [6,7]. Previous studies have confirmed the presence of both lectins and tannins in the aerial part of cassava [8,9].

The effectiveness of extracting proteins and phenolic compounds from plant tissues is influenced by various factors such as pH, temperature, time and the type of extracting agent, among others [10,11]. These factors can influence the intensity of the agglomeration phenomenon associated with the aforementioned active ingredients. Therefore, optimizing the extraction process is important in order to determine the best working conditions. In addition, the storage conditions of the extracts obtained play an important role in their stability [12].

In the case of lectins, the crude or purified extract can be tested in wastewater pollutant removal applications after detecting the agglomeration phenomenon and determining the affinity of the protein for a particular type of erythrocyte [13]. Once the activity has been detected, the crude or purified extract can be tested in wastewater pollutant removal applications [14]. Similarly, tannins can be used as an active ingredient in water treatment to remove different types of pollutants [15] and, in some cases, by subjecting them to physical-chemical transformations to improve treatment efficiency [16].

The ability of lectins and tannins to agglomerate has encouraged research into the use of these active ingredients in coagulation-flocculation processes, stages normally included in the primary treatment of wastewater [17]. In these processes, the natural compound, by means of transport phenomena in the diffuse layer, adsorption, bridge formation, sweeping and charge neutralization, destabilizes the colloidal and suspended particles and removes them from the treated effluent [18].

The use of natural organic coagulants appears to be a sustainable technology in wastewater treatment [12], becoming an alternative to the use of common coagulants, such as aluminum and iron salts, which are associated with various environmental problems [19]. It is well known that water quality parameters can significantly affect effluent treatment; in addition, coagulation process variables such as coagulant dosage and pH, among others, determine the effectiveness of the compound used in treatment [20].

Among the most problematic pollutants are dyes which, in addition to the intrinsic pollution they generate, reduce the penetration of light into bodies of water, affecting the amount of dissolved oxygen and thus hindering photosynthetic processes [21]. On the other hand, some dyes are toxic and mutagenic, and others, such as basic dyes, have the potential to release compounds classified as carcinogens [22].

In this context, this study aims to preliminarily evaluate the agglutination capacity of crude extracts obtained from the aerial part of cassava. Firstly, the hemagglutinating activity on different types of erythrocytes is evaluated, together with the optimization of the extraction and storage conditions for these extracts, in order to maximize the hemagglutinating activity. The extracts are then tested in decolorization trials of aqueous solutions of cationic dyes, also determining the most suitable conditions for removing the color.

2 OBJECTIVES

2.1 General Objective

To evaluate the agglutination phenomena of the aqueous and ethanolic extracts obtained from the aerial part of cassava by means of hemagglutinating activity tests and cationic dye removal tests in aqueous solutions.

2.2 Specific objectives

The specific objectives are:

a. To establish the optimum extraction and storage conditions for the crude and partially purified extracts, under which the highest hemagglutinating activities are obtained.

b. Evaluate the selectivity of the protein present in cassava leaves through hemagglutinating activity tests on different types of erythrocytes.

c. Select the most suitable raw material (leaves or stem) and the extracting agent that gives the best results (water or 70% ethanol) in the discoloration tests.

d. Determine the maximum color removal efficiency of aqueous solutions of cationic dyes, using the extracts as coagulating agents.

e. Characterize the protein, phenolic compound and tannin contents of the leaf and stem flours and relate the contents obtained to the agglomeration phenomena.

3 LITERATURE REVIEW

3.1 Cassava

Cassava *(Manihot Esculenta* Crantz*)* is a perennial, shrubby plant belonging to the Euphorbiaceae family. It originated in the Brazilian Amazon, a tropical region of South America, over 500 years ago and was taken to Asia and Africa in the 16th to 18th centuries [1]. It is easily adaptable to poor soils, highly resistant to pests or biological changes, has a low demand for nutrients and, in general, is highly tolerant of extreme climatic conditions. These characteristics facilitate its diffusion, making it a good option for farmers. Cassava is one of the main food sources in developing countries, where it is grown on small areas using low-level technologies [23,24]. Approximately 70 million people obtain more than 2100 $J \cdot d^{(-1)}$ of energy from cassava and more than 500 million consume quantities greater than 420 $J \cdot d^{(-1)}$ from the different products obtained from this plant [25].

According to the FAO [26], Brazil's cassava production in 2009 was 24.4 million tons, making it the world's second largest producer after Nigeria. In addition, per capita consumption is one of the highest (10.8 $kg \cdot hab^{-1} \cdot year^{-1}$) for cassava varieties, flour and starch [27]. Production is concentrated in the states of Parà, Bahia and Paranà, which account for 49% of the total cultivated in Brazil. All production is restricted to the domestic market and is far from participating in the international market, where Asian countries such as Thailand and Indonesia account for the largest percentage of exports [28]. These countries have high-tech physical and chemical processing processes and research centers where they develop new products from cassava.

The main industrial use of cassava is to use the tubers, the underground part of the plant, as food. There are two types of cassava: mansa and brava, the latter with linamarin concentrations above 50 $mg \cdot Kg^{-1}$ of tuber. Linamarin is a glycoside belonging to the plant's defense mechanism that is transformed into cyanide acid in the human body, causing alterations in the digestive system when consumed [29]. This type of cassava is normally used to make industrial flour. Table cassava is eaten boiled or fried in the form of various types of food [30].

The aerial part of the plant is treated as a by-product, and one of the uses is that of the leaves as a human food supplement or as animal feed due to their protein content. However, in most cases it is discarded, generating a considerable amount of waste [2]. In northeastern Brazil, cassava leaves form part of the so-called multimixture, which is provided in school meals and in the food basket for needy families [10].

3.1.1. Chemical composition of cassava aerial part

The aerial part of cassava includes the leaves, stems and stalks of the plant (Figure 1). Every year, 2.5 $t \cdot ha^{-1}$ of cassava leaves are produced and around 180,000 tons are wasted annually [31]. There is therefore a clear need to make use of this part of the plant in order to reduce the waste generated.

Figure 1: Cassava aerial part (leaves and stem)

Based on a total of 1.88 million hectares cultivated in Brazil in 2010 [28], considering an average weight of 0.45 kg of the aerial part per plant (one third of the total weight) and a planting density of 20000 plants$\cdot$ ha^{-1} [32], approximately 17 million tons are produced without practical use. Comparing the amount of leaves wasted annually and the tons generated by the loss of the aerial part, the stem of the plant represents the largest part of the waste generated.

Currently, chemical composition studies focus on the qualitative and quantitative determination of nutritional and anti-nutritional components, with the aim of making better use of the cassava plant as a food. Most of the characterization is done on the leaves due to their higher protein content, which compares to the protein content found in the stems [29]. *Table 1* shows the average composition of nutritional components in cassava leaves and Table 2 shows the average values of some anti-nutritional components.

Table 1. Average content of nutritional components in flour from 12-month-old cassava leaves, cultivar Ouro do Vale.

Constituent	Value[1]
Humidity (%)	9,2
Protein (g- 100g^{-1})	29,2
Vitamin C (mg$\cdot$100g^{-1})	64,1
Fiber (g$\cdot$100g^{-1})	24,1
β-carotene (mg$\cdot$100g^{-1})	124,2

[1]Protein, vitamin C, fiber and β-carotene values are presented on a dry basis. Source: [33]

Table 2. Average content of anti-nutritional components in 12-month-old cassava leaves, cultivar Ouro do Vale.

Constituent	Value[1]	Unit
Polyphenols	61,5	~-1 mg$\cdot$g^{-1}
Cyanide	11,29	mg$\cdot$100g^{-1}
Oxalates	2,48	g$\cdot$100g^{-1}
Saponins	1,74	g$\cdot$100g^{-1}

Source: [3]

[1]Values on a dry basis

The levels of the above components are related to the age of the plant, the degree of leafing and the cultivar of the plant. For each compound considered to be anti-nutritional or toxic, there is a plant age where the concentration will be higher or lower [32]. Along with their high protein and vitamin content, cassava leaves have acceptable levels of minerals such as calcium, iron and zinc [33,34].

Among the anti-nutritional components are polyphenols (which include tannins), substances with hydroxyls linked to aromatic rings, which reduce the digestibility and availability of amino acids such as lysine and impair protein utilization [9].

In addition to the substances shown in Table 2, cassava leaves contain glycoproteins called hemagglutinins or lectins which have the ability to bind to carbohydrates. These proteins can cause inconveniences in the utilization of nutrients or interfere with their absorption by the human body [35].

There is little information on the chemical composition of the stem. It regularly has a high fiber content and low moisture content compared to the leaves [36]. The protein content is around 5-6% [30], a much lower percentage than that of the leaves (Table 1). Because of this difference, the stem is not used in food supplements. In addition, the difficulty in grinding it [30] is another disadvantage for its use as food. Although they are mainly found in seeds, the presence of hemagglutinins has been proven in other vegetative tissues such as the stem [6,37]. There are no studies on the content of polyphenols in cassava stems.

Although there are substances that are harmful to the use of stems and leaves as food, these compounds can be used in other industrial applications [12,14]. In the specific case of this study, the research is aimed at evaluating the coagulating or agglutinating capacity of hemagglutinins and polyphenols present in crude extracts obtained from cassava stems and leaves.

3.2 Lectins

Lectins are glycoproteins with the ability to agglutinate erythrocytes and precipitate polysaccharides or carbohydrates with complex structures [4]. Lectins account for 2 to 15% of the total protein content in plants [38]; their functions in plants include the interaction and transport of carbohydrates and also participation in defense mechanisms against pathogenic microorganisms and insects [6]. Lectins are mostly found in plant seeds [14], but also in leaves, stems and roots, each of which has different structures and degrees of specificity [39].

Lectins often show selectivity in their interaction with sugars, forming bonds with specific molecules. For example, in experiments to characterize agglutination, it is common to observe differences in hemagglutination between different erythrocytes, which have surface carbohydrates present in blood cells, and even between different Iectins present in plants of the same family [38]. This agglutination process is reversible, leaving the structure of the carbohydrate to which the lectin binds unchanged [40].

The mechanism of action of lectins on sugars has been widely studied: the interaction occurs in the non-catalytic domains of the proteins where the reversible bond with the carbohydrate is formed; this bond is made by Van der Waals forces or by hydrogen bridges. Lectins often have two or more binding sites, which allows the formation of cross-links that can precipitate the bound compound [37]. Figure 2 shows the interactions between a lectin with two binding domains and erythrocytes.

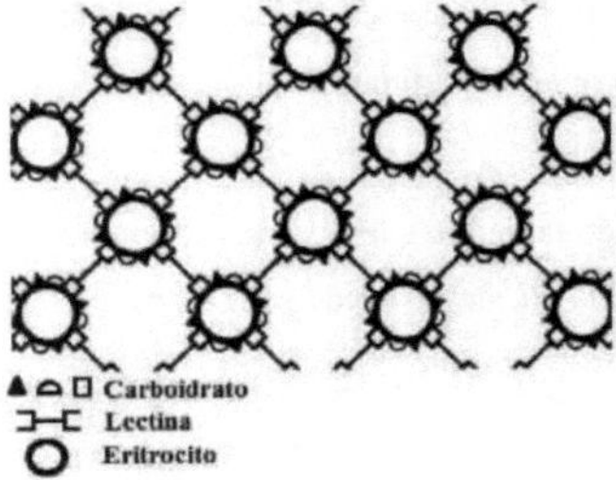

Figure 2 - Schematic representation of the lectin agglutination phenomenon [41].

Depending on their structure, the way in which binding occurs and their ability to precipitate molecules, lectins are classified as [6]:

- Merolectins: They have a binding site for carbohydrates. Due to their monovalent nature, they are unable to precipitate glycoconjugates; an example is the protein that binds to the chitin in the latex of the rubber tree *(Hevea Brasiliensis).*
- Hololectins: contain at least two identical or similar carbohydrate-binding sites. Due to their divalent nature, they bind and precipitate glycoconjugates.
- Chemolectins: They have one or more carbohydrate-binding sites and another site with catalytic activity (or other biological activity) that functions independently of the carbohydrate-binding site. Depending on the number of sites, they act as merolectins or hololectins.
- Superlectins: Have at least two different carbohydrate-binding sites.

The interaction with carbohydrates explains the toxic effect of lectins, because when they are ingested and pass through the digestive tract without being degraded, they can bind to the carbohydrates present in intestinal cells [41]. Thus, they can interfere with the absorption and use of nutrients, causing deterioration in some organs and generally affecting the body's metabolism [42].

Taking advantage of their structural characteristics, lectins are used in therapeutic applications such as antitumor agents in the detection and treatment of cancer [43], in the development of diagnostic tools in histology [37], as biosensors [44] and in general in the isolation of biomolecules such as polysaccharides [40] and enzymes [45].

The affinity of lectins for organic compounds makes it possible to propose the idea of removing specific substances from liquid effluents. The study of the affinity of the lectin obtained from *Moringa Oleifera* with water compounds for their removal from surface water is an example of this type of application [6].

3.2.1. Presence of lectins in cassava aerial parts

The presence of lectins in plant tissues is determined by evaluating their hemagglutinating activity (HA), which is defined as the minimum amount of protein per unit volume needed to agglutinate the erythrocytes present in a previously processed sample, evaluating serial dilutions on a base 2 basis.

Previous studies indicate the presence of hemagglutinins in cassava leaf concentrates from various cultivars. The intensity of the activity is inversely proportional to the age of the plant, with the highest AH values found at 12 months of age. In 15- and 17-month-old plants, the agglutination phenomenon is lower or non-existent. Maximum hemagglutinating activities have been detected up to the first [3] and second dilutions [8,13].

In the case of the stem, there are no previous studies on the possible presence of hemagglutinins. However, according to previous analyses, the percentage of protein decreases in flours that include both the leaves and the stem of the plant [46]; therefore, a decrease in HA is expected in extracts obtained from the stem. In addition, factors such as cultivar variety, soil fertility and water availability can affect the percentage of protein present in the aerial part of cassava [47].

3.2.2. Iectin extraction

After previous drying and conditioning steps, the extraction process commonly includes obtaining crude extracts, often aqueous or using saline solutions or buffers [6,48]. The extracting agent and the solid:liquid ratio are factors that significantly influence the amount of protein extracted. Pereira et al. [13] concluded that using distilled water in proportions of 1:20 (g of cassava leaves:ml of water^{-1}) gives better yields than extractions made with buffered saline solutions or in other proportions. Antov et al. [49] found higher protein contents in saline extractions than in aqueous extractions when working with bean seeds. In addition to these factors, there are other variables that influence protein extraction yield, such as pH, temperature, extraction time and stirring speed.

pH values close to 7 require a lower concentration of protein to cause hemagglutination; basic pHs decrease HA by approximately 50% while pHs below 4 often inhibit hemagglutinating activity [50].

Increases in temperature cause a decrease in HA, this decrease being largely dependent on the molecular weight of the lectin: low molecular weights help to maintain the stability and intensity of HA [50]. In some cases, an increase in temperature (above 40°C) causes a partial loss of HA, reaching a total loss at temperatures of 90°C, possibly due to protein denaturation processes [39]. It is therefore important to establish a suitable extraction temperature with viable operating conditions and optimized energy consumption.

The extraction time and stirring speed help to increase and improve contact between the solvent and the solid [13,51]. Although these conditions are not commonly optimized and are established as parameters, they can influence the final HA of the extract [10].

3.2.3. Partial purification of lectins

The isoelectric precipitation of lectins, and proteins in general, is influenced by their solubility, which can vary from practically insoluble (<10 μg· $^{mL(-1)}$) to very soluble (>300 μg· $^{mL(-1))}$. Critical factors affecting the solubility of a protein are pH, ionic strength, the nature of the ions, temperature and the polarity of the solvent. Proteins are generally less soluble at their isoelectric point where charge repulsion is lower. Protein fractionation is commonly done by adding ammonium sulphate, as this provides a stabilizing medium for the proteins [52], as well as being relatively inexpensive, pure and highly soluble [53].

Figure 3: Precipitation of protein with ammonium sulfate from crude extracts. Adapted from [54]

It is common to refer to the concentration of ammonium sulphate as the percentage of saturation. In general, the solubility of a protein decreases by a factor of 10 with a 6% increase in the amount of ammonium sulphate [54]. Figure 3 illustrates the influence of the amount of salt on the amount of protein precipitated from an extract.

Fractionation influences the purity and quantity of protein obtained. When the aim is to obtain the highest possible amount of protein, one should sacrifice obtaining a highly pure protein for greater recovery efficiency. On the other hand, if fractionation is the first step in obtaining a raw material, purity should prevail over the amount precipitated [54]. It is common to use successive fractionations to increase the percentage of saturation in order to increase the purity of the protein [13].

Subsequently, and depending on the final objective of using the protein, there are various purification methods such as ion exchange chromatography [55] and gel filtration [56], the basis of which is the variation in the charge, size and shape of the molecules [53]. A hypothetical example of the degree of purification is shown in Table 3; the specific activity column refers to the increase in a property, in this case HA, based on the activity of the crude extract.

Table 3. Summary of hypothetical protein purification

Fractionation	**Protein (mg)**	**Total activity (%)**	**Specific activity**	**Step yield (%)**	**Total yield (%)**
Crude extract	12000	100	1	75	100
Precipitation with ammonium sulphate (4550%)	1800	75	5	80	75
Ion exchange chromatography	240	60	30	75	60
Gel filtration	36	45	150	-	45
Pure standard	-	-	150	-	-

Source: [53]

Salt fractionation with ammonium sulphate is followed by exhaustive dialysis operations to remove the salts. At this stage, the appropriate choice of dialysis membrane directly influences the efficiency of the process [57]. A membrane with a very high exclusion limit can lead to the loss of low molecular weight proteins, affecting the yield [10]. Therefore, prior information on the molecular weight of the protein being studied is important.

Some works found in the literature present experimental studies with ground cassava leaves. Silva [10] obtained a protein recovery of 58% with precipitation in aqueous extract with ammonium sulphate at 80% saturation, with a consequent increase in AH of 30%. On the other hand, Pereira [58] worked with precipitation at different saturation percentages, obtaining the best protein yield at 50% saturation and the highest HA at saturation percentages of 25 and 50%; at values higher than 80%, no activity was detected.

In this study, precipitation with ammonium sulphate will be used to evaluate the effectiveness of the coagulation phenomenon of the crude extract, comparing the HA before and after fractionation.

3.2.4. Proteins as an active ingredient in water treatment

Once lectins have been detected in a plant tissue and their HA has been assessed, their behavior as coagulant molecules can be studied, either in the crude extract or using the partially or fully purified protein. The coagulating property of these molecules can be explained by the bridge formation model: the coagulation of negatively charged particles is the result of their binding to positively charged particles and the subsequent neutralization of the surface charge [59]. Post-neutralization adsorption phenomena have also been proposed as part of the particle agglomeration process [60]. In the case of *Moringa Oleifera*, there is debate about the composition of the active agent that causes coagulation. Some studies conclude that this is due to the presence of a cationic protein (lectin) [61]; others claim that the active ingredient is a low molecular weight polyelectrolyte other than a protein or polysaccharide [62].

Several studies have been carried out to evaluate the coagulating capacity of plant lectins as active ingredients: aqueous extracts of *Moringa Oleifera* seeds [61,63] and saline extracts of bean seeds [48,49]. Santos et al. [14] tested the lectin present in *Moringa Oleifera* seeds, stems and leaves and found it to have an acceptable performance in removing turbidity in simulated effluents in the laboratory.

Research into the coagulating activity of protein compounds in wastewater suggests that applying the product immediately after adequate purification can produce satisfactory results [48,64]. However, comparative studies of crude extracts and isolated lectins indicate the possibility of using the extract with an efficiency similar to that of the purified compound [14], and the use of the crude extract becomes a more economical alternative by eliminating processing costs.

3.3 Tannins

Tannins are secondary phenolic components of plants that are widely distributed in the plant kingdom [5]. These substances have no function in the primary metabolism of plants, but play various roles in protecting plants, for example from attack by diseases and herbivorous animals, and even act as toxic hormones, inhibiting the digestion of proteins. Tannins have a wide variety of molecular weights and complexity in their structure.

Their chemical composition is based on the presence of polycyclic aromatic compounds which give them the ability to bind to different types of molecules [7]. Its multiple hydroxyphenolic groups favor the formation of primary complexes with polysaccharides, metals, amino acids, proteins and other compounds [65].

Significant amounts of tannins are present in all parts of the plant. The concentration of tannins changes from one plant to another; in general, the percentage varies within a range of 1-5% [66]. The amount of tannins depends on the metabolic rate in the synthesis of phenolic compounds and the degree of polymerization of polyphenols, reaching higher percentages when high molecular weight components are formed, as occurs in seeds [67]. Tannins can be classified into two groups: proanthocyanidins (condensed) and the polyesters of gallic and hexahydroxydiphenic acids (hydrolysable, gallo and ellagitannins respectively) [68]. Condensed tannins are mainly present in the stems, leaves and stalks of plants, while hydrolysable tannins are commonly found in the leaves of plants and shrubs in tropical areas [5,69].

3.3.1. Condensed tannins

Commonly called proanthocyanidins or polyflavonoids, condensed tannins are common in some plants [5]: the fundamental structural unit of this group is the phenolic flavan-3-ol (catechin). Condensed tannins exist as complexes of oligomers (soluble in water) and polymers (insoluble in water) of flavonoid units linked by carbon-carbon bonds [5,66]. They are resistant to hydrolysis but, depending on their structure, can be soluble in aqueous solvents [70].

The analysis of condensed tannins is complex due to the diversity of structures present in this group of compounds. In summary, Figure 4 represents the basic structure of the repetitive unit, where the radicals R1, R2, and R3 can be OH groups which can be esterified, or they can be H groups. The different combinations of these three radicals determine the reactivity of the tannin. The presence of OH units in the rings can increase their ability to bind to protein-like molecules [71].

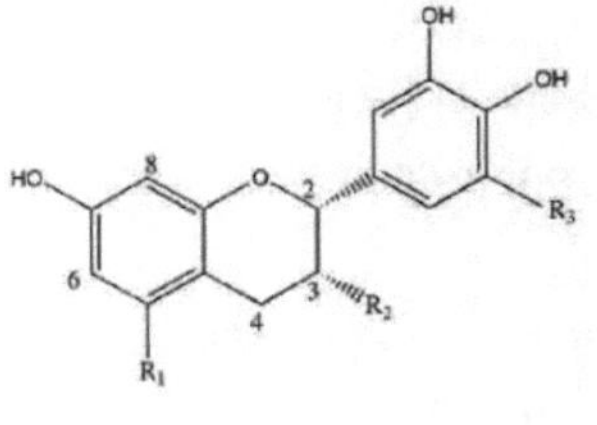

Figure 4: Repetitive basic unit present in condensed tannins.
Adapted from [71]

The polymerization process to form the condensable tannin structure includes an oxidizing bond between the flavonoid monomers (Figure 4) in positions 4, 6 and 8. The number of flavonoid units in the final polymer is in the range 2-17. Polymers with higher degrees of polymerization are insoluble. Figure 5 shows a model structure of a condensed tannin indicating the end point of the polymer [71].

The interactions of condensed tannins are based on hydrogen bridges and hydrophobic bonds. It is generally accepted that these interactions are strongly influenced by pH; the formation of complexes is favored at pHs greater than 3.5; below this value these complexes usually dissociate [5].

There are various techniques for analyzing tannins and total phenols in plants, such as the butanol-acid test and the Folin-Ciocalteu method, each of which has its advantages and disadvantages. Depending on the molecular structure, each phenolic component produces a different color intensity per unit mass and, therefore, the results obtained by colorimetric methods are used for semi-quantitative comparisons [72]. The generally accepted method for determining condensed tannins is colorimetric determination by the Butanol-HCl test, based on oxidative depolymerization which generates a red color characteristic of flavonoid molecules, the structural constituents of tannins. This procedure has limitations and considerations that must be taken into account when carrying out the analysis [71].

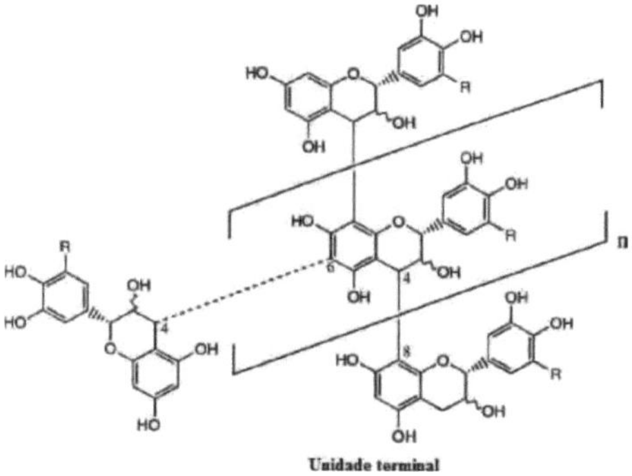

Figure 5 - Structural model of condensed tannins. Adapted from [71]

3.3.2. Hydrolysable tannins

Hydrolysable tannins are present in leaves, stems, bark, seeds and fruit. Some species can synthesize both ellagitannins and gallotannins and there are also plants that produce hydrolysable and condensed tannins or complexes with characteristics of both classes [72]. Unlike condensed tannins, hydrolysable tannins are easily degradable [73].

Hydrolysable tannins are made up of mixtures of simple phenols, such as ellagic acid, and esters of gallic acid with sugars, such as glucose [70]. Their ester-carboxyl bonds mean that they are quickly hydrolyzed under both basic and acidic conditions [74]. The basic structural unit is the polyol pentagalloylglucose (PGG) formed by the esterification of glucose by gallic and diglycolic acids (galotannins) or by hexadihydroxyphenic or ellagic acid (ellagitannins) [70]. The molecular structure of these acids is shown in Figure 6.

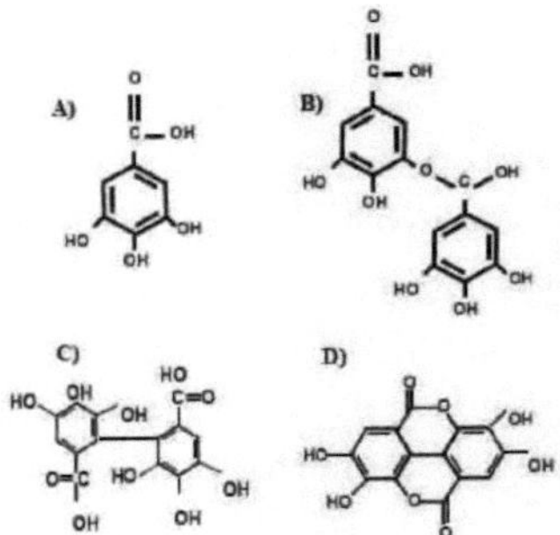

Figure 6: Molecular structures of gallic (A), diglylic (B), hexadihydroxyphenic (C) and ellagic (D) acids. Adapted from [70]

The most common of the gallotannins is tannic acid (Figure 7), usually composed of a core of 6 to 9 gallic acid units [70]. Simple gallotannin molecules are difficult to find in nature, but they tend to be the constituent units of commercially available tannic acids [72]. Figure 8 shows the molecular structure of an ellagitannin where the phenolic groups are molecules of hexahydroxydiphenic and ellagic acids.

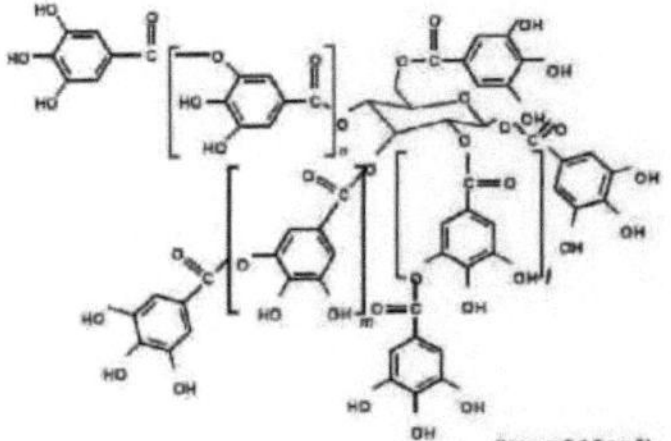

Figure 7. Chemical structure of a gallotannin (tannic acid) [70].

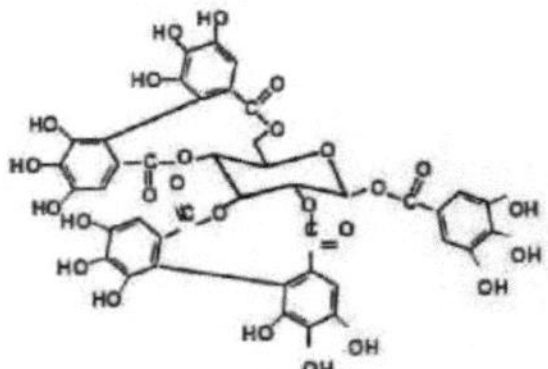

Figure 8. Chemical structure of an ellagitannin [70]

3.3.3. Presence of tannins in cassava aerial part

The presence of polyphenols in cassava leaves is well known, especially condensed tannins [34]. This presence limits the use of the aerial part of cassava as food due to the decreased digestibility caused by the interaction of tannins in the digestive system [75]. Marie Magdeleine et al. [76] present a characteristic phytochemical profile of cassava leaves (Table 4), indicating the presence of condensed tannins which make up a large percentage of the total polyphenols. Hydrolysable tannins were not detected.

Table 4: Phytochemical profile of cassava leaves

Component	Presence and intensity
Phenols	+
Tannins	++++
Hydrolysable tannins	--
Condensed tannins	++++
Flavonoids	+++
Anthocyanins	--
Proanthocyanidins	++++
Quinones	--
Sterols	++
Alkaloids	--
Coumarins	--
Anthracenes	--

Source: Adapted from [76]

Vanillin test studies confirm the presence of condensed tannins in methanolic and aqueous extracts of cassava leaves, in percentages of 10.0 and 25.0 % respectively. HPLC analyses indicate the presence of flavan-3-ol units and proanthocyanidins, structures typical of condensed tannins. Although aqueous extracts are more efficient, methanolic extracts have specific compounds that are not extracted with water [76,77].

Melo et al. [34] and Fasuyi and Aletor [2] found percentages of total phenolics in cassava leaf flour of 47.2 and 97.0 mg· $^{g(-1)}$ on a dry basis using ethanol (50%) and acetone (70%) respectively. Of the total polyphenols, there is a certain amount of tannins; concentrations in the range of 30-50 mg· $^{g(-1)}$ on a dry basis are considered typical in cassava leaves; however, there is a variation in the amounts found in different studies. While Awoyinka et al. [78] found 29.7 mg· $^{g(-1)}$ in a variety from Nigeria and Dung et al. [79] reported concentrations of 23 mg· $^{g(-1)}$, other studies have detected much lower amounts. Wanapat et al. [80] found values of 2.6 mg· $^{g(-1)}$ and Mondolot et al. [77] found a concentration of 1.83 mg· $^{g(-1)}$ expressed in catechin equivalents. The diversity of the methods used to extract and determine tannins may be the cause of these variations [77]. With the aim of increasing the digestibility of cassava leaf flour, Correa [9] tested different solvents to remove polyphenols; Table 5 shows the percentages obtained for the different solvents tested.

Table 5: Polyphenol content in cassava leaves before and after extraction

Solvent	Polyphenols (mg tannic acid:g dry matter)		Removal (%)
	Before	After	
Water	63,8	22,4	64,9
Ethanol (50%)	63,8	10,6	83,3
NH_4OH	63,8	3,7	94,2

Source: Adapted from [9]

It was previously stated that the age of the plant influences the presence of lectins in the leaves; in the case of polyphenols and therefore tannins, the percentage decreases with the age of the tree. Wobeto et al. [3] determined polyphenol contents in a range of 70-106 mg· $^{g(-)}$1 in leaves collected at 17 months, while in leaves collected at 12 months the contents were between 43 and 62 mg· $^{g(-1)}$. In a similar study, Teo et al. [81] found tannin concentrations of 60 and 120 mg· $^{g(-1)}$ for leaves harvested at 6 and 12 months respectively.

There are studies on the presence of polyphenols in mixtures of cassava leaves, stems and stalks. Thang et al. [82] and Wanapat et al. [83] determined levels of 12.0 and 40.0 mg· $^{g(-1)}$ on a dry basis for mixtures of plants collected three months after planting. In these studies there was no data on the proportion of stems used in the mixtures. Dung et al. [79] reported condensed tannin contents of around 20-30 mg· $^{g(-1)}$ in mixtures containing a small proportion of ground stems and stalks. Hue et al. [84] found that the percentage of tannins in mixtures of freshly ground leaves and stems is higher than that found in mixtures stored for long periods.

3.3.4. Tannin extraction

The extraction of tannins is influenced by their chemical nature, the method used, the particle size of the solid, storage conditions and the presence of interfering substances. Tannins may be forming complexes with carbohydrates, proteins or other compounds, making it difficult for them to be soluble and, for this reason, additional steps may be necessary to guarantee total removal [85]. For this, solid phase extractions are commonly used [86].

In general, there are simple solid-solvent contact processes [65,87], modifications applying reflux (Soxhlet method) [11] and procedures that seek to increase contact, such as microwaves and ultrasound [88].

One of the determining factors during the extraction process is the choice of solvent. Correa et al. [9] found effective removal with saline ammonium hydroxide solutions; Teo et al. [81] determined that the addition of sodium sulfite to aqueous solutions increases the amount of tannins extracted. However, the use of these salts can generate residual toxicity, as well as being expensive and difficult to handle.

The polarity of the solvent and the solvent-to-solid ratio are factors that strongly influence the choice of extracting agent [11]. Acetone-water and methanol-water mixtures are highly efficient at extracting tannins, especially condensed tannins [87,89]. When aiming for environmentally friendly applications, these solvents generate vapors and are dangerous to handle and toxic. Ethanol has been used successfully [11,90], making it an easily accessible and less expensive alternative, and its volatility can be used to recover the solvent and concentrate the extract [81].

Chavan et al. [67] found the best yields in the extraction of condensed tannins from *Lathyrus Maritimus L.A.* using acidified aqueous solutions of 70% acetone. Table 6 summarizes the results obtained in the extraction of tannins from the plant *Phyllanthus niruri Linn*, comparing the efficiency achieved for each solvent.

It is common to use successive extractions to increase the amount of tannins extracted. Naczk and Shahidi [86] indicate that after three successive extractions the increase in phenolic compounds in the crude extract becomes imperceptible.

The prior drying of the samples to be extracted also influences the percentage of tannins extracted. It may seem that freeze-drying is the most suitable method for preserving the sample; however, although this process maintains the polyphenol composition, there is a decrease in the ability to extract the tannins; in addition, the energy costs associated with this process are high. Drying with an increase in temperature is the best alternative, however, few quantities of tannins are extracted in samples dried at high temperatures due to the alteration of the molecular structure of the polyphenols. Drying at room temperature is therefore preferred

[87].

Table 6. Effect of solvents on tannin extraction

Solvent[1]	**Snyder polarity index**[2]	**Boiling point (°C)**	**Tannins (mg·g(-1) of sample-1)**[3]
Organic			
n-Hexane	0,1	69	18
Petroleum ether	0,1	60	22
Dichloromethane	3,4	40	40
Chloroform	4,1	61	97
Acetone	5,4	56	39
Ethanol	5,2	78	116
Methanol	6,6	65	146
Acuoso[4]			
Acetone 70%	6,5	84	185
70% ethanol	8,2	90	208
Ethanol 50%	7,9	94	225
Ethanol 30 %	7,1	97	264
Ethanol 20%	6,3	98	271
Water	9,0	100	262
Water (treated with hexane)		100	235

Source: Adapted from [11]

[1] Extracts obtained by the Soxhlet method, sample-solvent ratio 1g:25ml, temperature: 25°C, time: 3 hours

[2] An increase in the index indicates greater polarity.

[3] Tannins expressed in mg of tannic acid.

[4] The percentage indicates the part composed of the organic solvent.

Some studies indicate that long extraction times favor the percentage of tannins extracted [11], while others claim that the yield is independent of the time the solids are exposed to the extracting agent [91] or, conversely, prolonged extractions can cause the compounds to oxidize. Extraction times ranging from 1 minute to 72 hours have been reported in the literature [86]. Depending on the purpose of the extraction, it is common to add antioxidant compounds such as ascorbic acid in concentrations of around 0.01% [65,92].

Agitation during extraction helps to increase removal efficiency by increasing solvent-solid contact [81]. Magnetic stirring, mechanical stirring, homogenizers, ultrasonic microwaves and sonication equipment can be used [86,88].

The recovery of tannins is also affected by the sample:solvent ratio. Common ratios vary between 1:5 and 1:20 (m:v).

Naczk and Shahidi [86] increased the extraction efficiency of canola tannins by 30% from 1:5 to 1:10 with 70% acetone [86].

Common pH values for cassava leaf extracts (aqueous or ethanolic) are in the order of 5-6 and in some cases reach neutral pH values depending on the characteristics of the polyphenols soluble in the extracting agent. [9].

In subsequent processes, in order to make the extracts applicable, to characterize them or to recover the solvents used, evaporation-concentration operations are carried out, taking advantage of the difference in the boiling points of the solvents [11,67,87,91].

3.3.5. Tannins as an active ingredient in water treatment

In general, synthetic polyelectrolytes (cationic, anionic and non-anionic) and natural polyelectrolytes, such as guar gum, starches and proteins, have been successfully applied as coagulants in water. Similarly, tannins can be used for this purpose due to the presence of carboxyl and hydroxyl groups in their structures [93].

There are several tannin-based products marketed as flocculants or coagulants. The species *Acacia mearnsii*, Castanha (tropical), *Quercus ilex*, sùber and robur (non-tropical) are examples of sources of tannins used in water treatment [15].

Tannin molecules are anionic in nature due to the presence of phenolic groups. Figure 9 is a schematic representation of the basic structure of a tannin in aqueous solution and the possible molecular interactions that lead to coagulation [94].

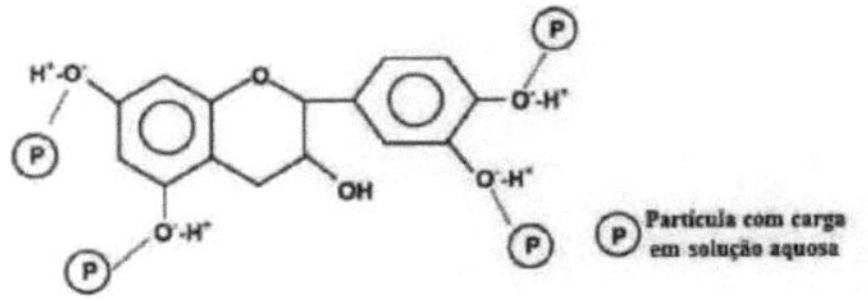

Figure 9. Representation of the basic structure of tannin in aqueous solution and its possible molecular interactions. Adapted from [94]

The effectiveness of tannin as an active ingredient depends on the available binding sites in the molecular structure and the degree of modification.

that they may undergo in the extraction and purification processes [19]. In the literature there are processes for the physical-chemical transformation of tannin mixtures based on the Mannich reaction, many of which are protected by patents [95,96,97]. This reaction, commonly known as cationization, is defined as the chemical procedure that confers a cationic character to the organic matrix of tannins through the addition of quaternary ammonium salts. These modifications give tannins an amphoteric nature due to the presence of amino groups (cationic) in the monomeric units and the phenolic groups (anionic) of the tannin [94].

The coagulating capacity of tannins increases by destabilizing colloids in solution, without altering the properties (solubility, stability) of the natural compound. Destabilization and subsequent sedimentation cause the removal of a wide variety of compounds, mainly anionic dyes and surfactants [98].

The use of natural extracts without exhaustive purification processes requires that discoloration tests be conducted with extracts that are not stored for long periods or in conditions that favor degradation, oxidative processes or denaturation. Jeon et al. [12] used polyphenolic extracts from grape seeds up to a maximum of 24 hours after being extracted and stored at 4°C.

3.4 Coagulation - flocculation

Water is an essential resource for human life. It is not found in its pure state in nature, and its presence is accompanied by substances in solution or suspension that can alter its characteristics and cause problems for health, the economy or the environment. In addition to the substances that naturally make it up, water contains substances from the soil or from various agricultural or industrial activities. Effluent treatment is therefore essential [6]. A general scheme for wastewater treatment involves three main stages:

- A primary treatment or pre-treatment using physical, chemical and mechanical methods.
- A secondary treatment or purification step applying chemical or biological methodologies.
- Treatment of the sludge or sub-residues formed.

Coagulation-flocculation processes are often applied in the primary stages of water treatment (in some cases they are also used in secondary and tertiary treatments). Coagulation consists of combining insoluble particles or dissolved organic matter into aggregated compounds, facilitating their removal in subsequent sedimentation, flotation or filtration stages. This process usually begins with the dispersion of a coagulating agent that destabilizes the colloidal particles, promoting the formation of microflocs. The aggregation of several microflocs, by the addition of the flocculating agent, leads to the formation of larger agglomerates which can be easily separated by mechanical operations [99].

The industrial process begins with the addition of the coagulant to the effluent in a rapid mixing chamber in order to quickly distribute the chemical throughout the body of water (coagulation). Subsequently, the body of water is subjected to slow agitation where larger flocs are formed (flocculation). Depending on the coagulating agent, the flocculation process takes place at the moment of addition, or another chemical agent is required. Both processes (coagulation and flocculation) are necessary to remove pollutants [100]. Finally, the effluent is taken to sludge sedimentation tanks, which are then taken to tertiary treatments. Figure 10 shows a common coagulation-flocculation scheme; the retention times at each stage depend on the volume of the unit used and the flow rate of the effluent.

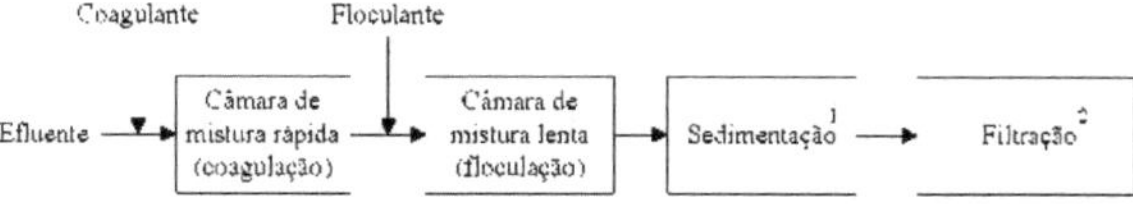

Figure 10. Traditional diagram of the coagulation-flocculation process in water treatment. Adapted from [17].

[1] Or alternative processes for removing settled solids.

[2] Or alternative suspended solids processes.

Colloids are stable in solution and, in general, the factors that stabilize them prevail over those that destabilize them. Stabilizing factors include repulsive forces, such as electrostatics and hydration. Conversely, attractive forces (such as gravity, Brownian motion and Van der Waals forces) cause destabilization. The principle of coagulation-flocculation can be explained by different mechanisms, all of which have in common the alteration of ionic forces in the medium. The literature proposes four destabilization mechanisms:

- Diffuse layer compression: Addition of oppositely charged ions to the colloids increases the total number of ions in the diffuse layer which, in order to remain electrically neutral, reduces its volume. In this way, Van der Waals forces destroy electrostatic stability [18]. In this mechanism, the quantity of electrolytes is independent of the concentration of the pollutant; moreover, there is no re-stabilization of the colloidal particles [22].
- Adsorption and bridge formation: A common phenomenon for organic compounds; the ionizable sites of the long molecular chains bind with the pollutant according to the character of the coagulant, which can be cationic, anionic or amphoteric.
- Adsorption and neutralization of charges: The coagulant molecule adsorbs onto the surface of the colloidal particle and then, because it has opposite charges, destabilizes the colloids. This phenomenon includes covalent interactions, hydrogen bridges, ion exchange and coordination reactions.
- Sweeping: The colloidal particles are surrounded by the precipitates formed. This mechanism is described in industrial processes where aluminum and iron salts are used, and the hydroxides formed are used [22].

Coagulant polymers are associated with adsorption, bridge formation and charge neutralization mechanisms, and constitute a starting point for proposing the mechanisms that prevail in natural coagulants [94]. Figure 11 schematically shows the coagulation-flocculation process using organic compounds, where there is a succession of elementary processes that occur simultaneously and according to the flow conditions in the medium (usually turbulent) [101]. The steps proposed in the diagram are:

i. Dilution of the flocculant in a homogeneous solution.
ii. Collision between colloidal particles.
iii. Transport of the added organic compound to the surface of the colloidal particles.
iv. Reconformation of the coagulant adsorbed on the surface of colloidal particles.
v. Formation of bonds (bridges) between colloidal particles.
vi. Rearrangement and disruption of the structure of the flakes formed.

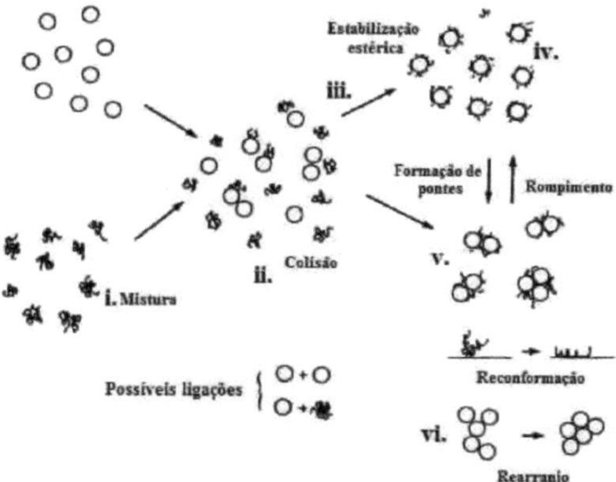

Figure 11. Schematic diagram of the coagulation-flocculation process using organic compounds. Adapted from [101]

There are two main classes of materials used in coagulation-flocculation processes:

- Organic and inorganic coagulants including mineral additives (lime, calcium, salts, etc.), hydrolyzed metal salts (aluminium sulphate, ferric chloride), prehydrolyzed metals (polychloride) and polyelectrolytes [102].
- Organic flocculants including cationic and anionic polyelectrolytes, nonionic polymers, amphoteric compounds and natural coagulants and flocculants (tannins, proteins, starches, alginates, etc.) [102].

3.4.1. Factors influencing coagulation-flocculation

It is well known that water quality parameters such as pH, organic matter concentration and temperature can significantly affect the treatment of an effluent. In addition, variables such as coagulant dosage, pH and organic load, among others, determine the effectiveness of the compound used in coagulation treatment [20].

The coagulation process can result in the formation of small, fragile flocs that can be broken up by physical forces. It is therefore important to provide the process with adequate flocculating capacity and rapid sedimentation of the flocs formed. An alternative is the use of adjuvants or additives to facilitate bonding and to agglomerate the flocs formed by the coagulant. Additives, many of which are polymeric (natural or synthetic), are characterized by their ionic nature: cationic, anionic or non-ionic. These compounds can act by bridge-forming mechanisms or by neutralizing charges [99].

A brief description of the factors that are commonly studied in the evaluation of any type of coagulant is presented below, emphasizing the natural active ingredients and how they affect or influence the final effectiveness of the tested substance.

3.4.1.1. pH

The performance of polyelectrolytes in coagulation-flocculation processes is mainly governed by pH, which is affected both by the charge given by the polyelectrolyte when it is added and by the solubility of the compounds [103]. In the addition of inorganic coagulants, such as aluminum and iron salts, changes in pH are important factors for effective treatment. In the case of natural coagulants, the pH ranges in which there is considerable effectiveness are greater than those of inorganic coagulants, allowing for a more flexible process without abrupt variations in the initial parameters of the process. When considering natural coagulants, the effect of pH is most noticeable in those containing tannins as their active ingredient [63]. When the essential component of the coagulant is a protein compound, the ionic force generated by the change in pH does not affect the conformation of the proteins, keeping the coagulant activity stable up to pH values of 4 to 9; outside these values, protein denaturation phenomena can occur.

The addition of coagulant compounds can lead to changes in the initial pH of an effluent: the use of inorganic salts generates a more marked decrease than in the case of natural coagulants. Ndabigengesere et al. [61], starting from synthetic effluents (water + kaolin) with an initial pH of 7, reached pHs of 4.2; however, when working with a natural coagulant (*Moringa oleifera* seed extract), the decrease only reached a final pH of 6.5.

Determining the optimum pH for natural coagulants can be difficult, even though these compounds do not significantly affect the pH of the effluent. In fact, the literature presents optimum pH values that are either acidic [104], neutral [64] or basic [105].

3.4.1.2. Temperature

In general, crude extracts obtained from natural raw materials and subjected to high temperatures do not reduce their coagulant activity, or their effect is much less than the effect of variables such as pH, the dosage of the coagulant or the initial concentration of the pollutant [105]. Santos et al. [14] found that a protein fraction purified from *Moringa oleifera* seeds was stable up to 100°C. Coagulation-flocculation operations are normally carried out at temperatures below 40°C, ensuring that the active ingredients present in the coagulant used are not degraded [15]. Beltran-Heredia et al. [63] concluded that when working at temperatures up to 40°C with *Moringa oleifera* extracts in the removal of dyes, the percentage of color removal does not undergo major changes. Some research has shown proteins to be more resistant, especially when purified [14].

The study temperature is associated with the actual conditions of the effluent in the process or in the environment where it is collected. For example, in the case of surface water, climatic fluctuations mean that coagulation studies are carried out at temperatures between 10 and 40°C [106]; this range accounts for a large percentage of industrial and domestic effluents.

3.4.1.3. Initial pollutant concentration

The initial concentration of the pollutant directly affects the required dosage of coagulant, influencing the final percentage of removal. There is no common consensus regarding the proportionality of the effect of this variable. Higher efficiencies have been reported at low dye concentrations, where the maximum percentage of removal is reached more quickly than in effluents with high concentrations [107]. Other authors indicate higher removal percentages when the concentration of the pollutant is high [104]. The cause of this difference is explained by the greater number of molecules available for collision with the molecular chains of the coagulants.

3.4.1.4. Agitation

Coagulation-flocculation begins with rapid mixing of the effluent with the coagulant, flocculant or both. The high speed of agitation distributes the compound (inorganic salt, synthetic or natural polymer) throughout the water mass. Subsequently, in a slow phase, flocs form and the molecules agglomerate, allowing the pollutant to sediment and then be removed [108].

The optimization of these stages depends on establishing a stirring speed, in the fast stage, which produces a turbulent flow so as to form a homogeneous mixture of particles, without the formation of flocs. In slow mixing, the aim is to encourage particle collisions and the formation of bridges, in order to achieve progressive floc formation [101]. The wrong choice of agitation times in the fast and slow stages causes problems such as the formation of fine flocs with no sedimentation capacity; on the other hand, optimizing these times can bring benefits such as a reduction in the dosage of coagulants; these advantages and disadvantages are more visible when working with low dosages of coagulant [99,106]. It is advisable to guarantee times equal to or greater than those considered optimal; insufficient times will cause a decrease in removal, with poor use of the coagulant's capacity [17]. Times of rapid agitation between 1 and 10 minutes and slow steps of 30 minutes are commonly used in coagulation tests [22,62,64].

Stirring speeds in the fast mixing stage are around 150-300 rpm. In the slow stage, common values are between 30 and 60 rpm [12,61,107]. There is no significant relationship between the speeds and times of the two stages.

3.4.1.5. Coagulant dosage

The formation of bridges in coagulation-flocculation depends on there being enough free sites on the pollutant molecule to allow the long chains of the organic coagulant to adsorb. An overdose of the coagulant can lead to a very high number of bonds on the surface of the pollutant, so that the re-stabilization of the colloids is favoured, preventing aggregation and subsequent floc formation (Figure 12) [102]. This excessive dosage can also increase the turbidity of the water due to the reversal of the colloidal load caused by the large quantity of adsorbed molecules; in addition, in the case of natural coagulants, the unbound protein remains in the solution, increasing the organic load of the effluent [61]. On the other hand, an insufficient dosage does not allow bridges to form and hinders pollutant-coagulant contact. These considerations lead to the idea of an optimum dosage, which is found at a value lower than the complete saturation of the surface by the coagulant; thus, it is proposed that the optimum dosage is directly proportional to the surface area of the pollutant particles [102].

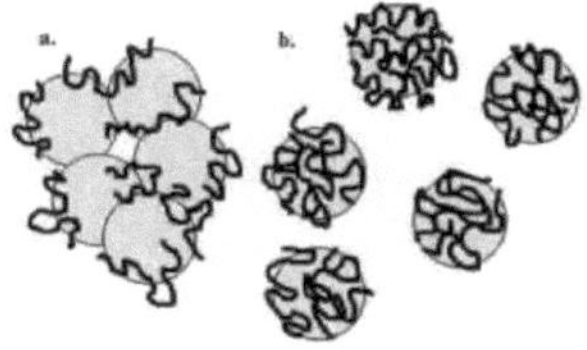

Figure 12. Schematic diagram of a) bridge formation and

flocculation and b) re-stabilization of colloids [102].

3.4.1.6. Addition of alkalizing agent

To facilitate the process of agglomeration of particles that could sediment, the alkalinity of the medium must be guaranteed, either by the original conditions of the effluent or by the addition of coagulation aids that provide a basic character to the medium. Compounds such as sodium carbonate and sodium and calcium hydroxides are commonly used in these treatments [22]. The choice of alkalizing agent is governed firstly by its effectiveness and secondly by its cost. In addition, it is important to assess the possible presence of residual ions (mainly carbonates) that could cause precipitation and fouling problems in industrial-scale equipment [109].

3.5 Natural coagulants

Aluminum and iron salts are the most commonly used coagulants in water and wastewater treatment. However, researchers have pointed out serious drawbacks related to the use of these salts, such as Alzheimer's disease, which is associated with residual amounts of aluminum in treated water [19]. In addition, disposal of the sludge generated during treatment is difficult due to the large amount of inorganic material present [49] and the reaction of aluminum salts with the natural alkalinity of the water, which leads to a decrease in pH [20]. To date, various coagulants such as ferric salts and polychlorides have been suggested as alternatives to aluminum salts

[102] . Although these materials develop properly in coagulation, positive consequences for health and the environment are not guaranteed.

Another type of coagulant used is synthetic polymers, which can be cationic or anionic. One of the problems presented by the use of these compounds is the presence of toxic monomers in the sludge formed, which are related to neuronal problems and carcinogenic properties [49].

The use of natural organic coagulants then appears as a sustainable alternative for wastewater treatment [12]. They have characteristics such as high charge density, long molecular chains, bridge formation and precipitation at neutral or alkaline pH, especially [99]. Replacing commonly used chemicals with natural coagulants reduces the costs of dosing coagulants and flocculants, and generates sedimentation sludge with greater biodegradability, smaller volumes (20-30% reduction), greater handling facilities and no potentially toxic substances [110]. In rural or remote areas, natural coagulants obtained from local plants are used for domestic purposes, even before the use of chemical coagulants, taking advantage of the low costs of obtaining, processing and availability [20,49].

The active ingredients, particularly protein and phenolic compounds, are extracted from different parts of plants: bark, stalks, leaves, fruit, flowers and even the roots [111]. Today, experiments with coagulants obtained from various plant species have been reported in the literature: beans (*Phaseolus Vulgaris*) [48], *Moringa Oleifera* [59,60,62,64], Opuntia spp [20], *Strychnos Potatorum* L. [94,112], *Cassia javahikai* [113], Okra (*Abelmoschus esculentus)* [114], Quebracho *(Schinopsis balansae)* [106], cassava starch [115], Acacia negra *(Acacia mearnsii)* [116], guar gum *(Fabaceae)* [117], *Jathropa curcas,* Chestnut *(Castanea sativa)*

[100], Maize *(Zeemays)* [118] and other agricultural materials [119].

In the industry there are already several brands marketed as coagulants or flocculants based on natural active ingredients. Products such as *Acquapol C-1*, *Acquapol S5T* and *Tanfloc*, obtained from *Acacia mearnsii* and marketed in Brazil, and *Silvafloc* derived from Quebracho *(Schiinopsis balansae)* and distributed in Italy, are examples of coagulants with tannins as their active ingredient [100].

The physico-chemical characteristics of natural coagulants open up the possibility of testing them with various pollutants, both in the form of particulate matter and dissolved compounds [99]. Table 7 shows a summary of some of the effluents *already* tested.

Table 7. Examples of effluents treated by coagulation-flocculation using compounds of natural origin

Effluent	Reference
Anionic surfactants	[15,116]
Inorganic suspensions (bentonite, kaolinite)	[14,49,61,62,64]
Effluents containing dyes	[12,63,100,107]
Mine effluents	[110]
Effluent containing metal ions	[112,113]
Effluents containing humic substances	[114]
Well water	[115,117]
Surface water	[100,106]
Water-oil emulsions	[110,120]
Domestic wastewater	[100]
Drinking water treatment	[93,121]

Natural coagulants can be used as primary coagulants to replace chemical or synthetic coagulants or flocculants, or even as adjuncts to them [49]. The efficiency of these compounds is influenced by the degree of purification. Various techniques are used to isolate the active ingredients, such as ion exchange fractionation, dialysis and lipid elimination [48,62].

The processes involved in obtaining natural coagulants from plants can be divided into three main stages (Figure 13):

- Drying and pulverizing the plant parts: the products obtained at this stage not only contain the active agents; other organic compounds may be present which can affect the effluent load and therefore the performance of the coagulant. However, their use without additional purification reduces treatment costs [122].

- Extraction: The extraction of active ingredients can be carried out using different solvents (organic, aqueous, salt solutions) which can influence the chemical structures extracted and the electrostatic properties of the extract obtained [122].

- Tertiary stages, rarely used in industry, and aimed at research to determine and characterize agents

active ingredients. These include freeze-drying, dialysis and ion exchange processes [61,62,123]. These procedures increase the total cost of processing and are only used industrially in a few cases, such as for *Moringa Oleifera* [94].

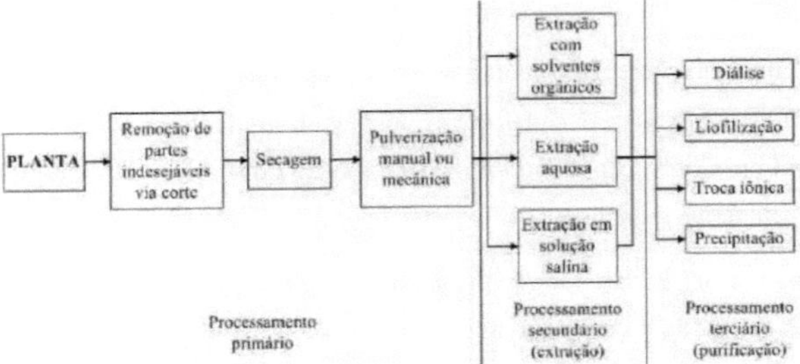

Figure 13. General processing steps in the preparation of natural coagulants obtained from plants. Adapted from [94]

The choice of extracting agent should favor the coagulating capacity without affecting the cost of the operation and without generating new contaminants in the effluent. Okuda [62] indicates that the coagulant extracted from *Moringa Oleifera* seeds with a saline solution of NaCl has a capacity 7 to 8 times greater than that extracted with water.

The use of natural coagulants brings with it variations in the effluent's chemical oxygen demand (COD), due to the presence of biomolecules that can increase the organic load [123]. Ndabigengesere et al. [64] and Anastasakis et al. [114] obtained increases of up to 50% in the initial COD value of the effluent. This change can exceed the chlorine demand and, as a consequence, favor the formation of trihalomethanes during the disinfection process. Purification of the crude extracts allows the effect on the increase in COD to be less [123]. Antov et al. [48] purified the protein obtained from bean seeds and obtained an organic load 15 times lower in the effluent treated with the purified extract than that obtained when treating the effluent with the crude extract.

An important aspect in the effectiveness of natural coagulants is their storage, either of the solids collected or of the extracts produced. Katayon et al. [124] found a decrease in the coagulant activity of M. oleifera seeds as the storage time of the raw material increased.

3.6 Dye removal

Large quantities of dyes are dumped into aquifer effluents every year. In addition to the intrinsic contamination they generate, their presence reduces the penetration of light, affecting the amount of dissolved oxygen more critically than other types of contaminants, as they hinder the photosynthetic process carried out by aquatic flora and alter the respiration processes of fauna [21]. Some dyes are toxic and mutagenic; others, such as basic dyes, have the potential to release compounds classified as carcinogenic due to the presence of structures such as benzidine [22].

The classification of dyes is based on various criteria, such as chemical structure or acidic or basic nature, among others [16]. One classification accepted worldwide is the Colour Index (CI), which is published by the SDC (Society of Dyers and Colourists) together with the AATCC (American Association of Textile Chemists and Colorists). In this database, dyes and pigments are grouped by color ranges with similar chemical structures [22]. Almost two thirds of organic dyes are azo compounds, followed in percentage by anthraquinones (15%).

Table 8 shows the categories established by the IC for the dyes tested in this work, which are classified according to their main molecular structure and illustrated in Figure 14. This structure is regularly composed of aromatic rings and an unsaturated functional unit where the color is developed (chromophore). In addition, there may be additional functional groups (-OH, -NH2, - N(CH3)2, etc.) attached to the rings, called auxochromes [125].

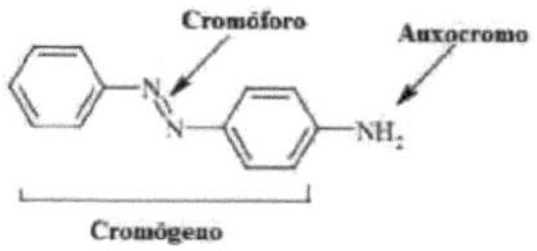

Figure 14. Example of the structure of a dye (4-aminoazobenzene - Strong yellow) [125]

It can be seen in Table 8 that for each molecular structure there can be several modes of application, which essentially depend on the surface or material where the dye is applied. Table 9 summarizes the main fields of use for dyes, listed according to the application method used for each dye; this classification is widely used in industry.

Table 8. Classification of dyes and pigments according to color index

Basic molecular structure	CI code	Functional group	Example	Classification by application modes
Monoazo	11000 19999	-N=N-	Blue Maxilon GRL 300%	Acids, direct, disperse, basic, mordant, reactive.
Trialrilmethane	42000 44999	O Qu	Malachite green	Acids, basics, mordants.
Thiazine	52000 52999	0	Methylene blue	Basics, jaws.

Source: Adapted from [21,22,125]

Over time, researchers have worked on developing techniques to remove dyes from wastewater and different procedures have been proposed: adsorption onto materials such as activated carbon and biosolids [21], physical or chemical degradation, Fenton oxidation, electrochemical degradation, ozonation, etc. [63].

Table 9. Fields of use of dyes according to their method application

Method of application	Main uses
To the vat	Natural and artificial fibers
In ink	Natural fibers
Acids	Food, leather, natural and synthetic fibers, la and
Sulphur	paper
Basics	Natural fibers (cellulose)
Direct	Leather, synthetic fibers, wool, wood, paper
Scattered	Leather, natural and artificial fibers, paper
Jaws	Natural and synthetic (acrylic) fibers
Reactive	Anodized aluminum, wool, natural and synthetic
Solvents	fibers
Pre-metalized	Leather, natural and artificial fibers, paper

Optical brighteners	Waxes, cosmetics, wood, plastics, varnishes, paints Protein fibers and polyamides Detergents, natural, artificial and synthetic fibers, plastics, soaps, paints, paper

Source: [21]

In coagulation-flocculation processes, different types of natural coagulants have been successfully used as primary agents to remove dyes. Table 10 presents a summary of this research, emphasizing the type of dye studied and its basic molecular structure. There are also studies where coagulants are used as secondary coagulants or as flocculants to reduce the use of the main coagulant, which is usually an inorganic salt [126].

Table 10. Coagulation-flocculation studies for dye removal reported in the literature using natural coagulants

Dye	Class	Coagulant	% removal[1]	Reference
Alizarin violet	Anthraquinone	a. *Moringa Oleifera* Extract	90	[98,100]
		b. ACQUAPOL C- 12	55	
		c. TANFLOC[2]	55	
		d. SILVAFLOC[2]	40	
		e. ACQUAPOL SST[2]	65	
		f. Chitosan[3]	20	
		g. Starch	10	
Methylene blue	Thiazine	a. *Moringa Oleifera* Extract	0	[16,63]
		b. ACQUAPOL C- 1	0	
Indigo carmine	Indigo	a. *Moringa Oleifera* Extract	55	[16,63]
		b. ACQUAPOL C- 1	8	
Palatine Black	Monoazo	a. *Moringa Oleifera* Extract	100	[16,98]
		b. ACQUAPOL C- 1	80	
Blue Sky Chicago 66	Diazo	a. *Moringa Oleifera* Extract	100	[16,63]
		b. ACQUAPOL C- 1	70	
Ericromocyanin R	Triarylmethane	a. *Moringa Oleifera* Extract	40	[16,63]
		b. ACQUAPOL C- 1	10	
Acid red 88	Monoazo	a. *Moringa Oleifera* Extract	100	[16,63]
		b. ACQUAPOL C- 1	75	
Diazo red congo	Diazo	a. *Moringa Oleifera* extract b. Corn extract c. Chitosan	70 45 65	[118]
Acid black 1	Diazo	Chitosan	82	[107]
Acid violet 5	Monoazo	Chitosan	83	[107]
Reactive black 5	Diazo	Chitosan	85	[107]
Triarylmethane green Green	Triarylmethane	Snail shell extract[4]	55	[104]

Malachite				
Orange direct 26	Diazo	Ipomoeae dasysperma seed extract[5]	50	[113]
Acid red Diazo 114		Seed gum extract extract Ipomoea dasysperma	30	[113]
Procion blue Anthraquinone brilliant		Seed gum extract extract Ipomoea dasysperma	20	[113]
Triarylmethane crystal violet		Grape seed extract	75	[12]
Triarylmethane Triarylmethane malachite		Grape seed extract	60	[12]
Diazo yellow gold		*Tamarindus indica* mucilage	60	[127]
Direct red Monoazo		*Tamarindus indica* mucilage	25	[127]

[1] Maximum percentage removal achieved for each study as a function of different variables (pH, coagulant dosage, dye concentration, temperature, etc.).

[2] Coagulants based on tannins extracted from *Acacia mearnsii* (TANFLOC, ACQUAPOL C-1, ACQUAPOL SST) and *Quebracho* (SILVAFLOC)

[3] Cationic polysaccharide of animal origin, found in the exoskeleton of crustaceans

[4] Of animal origin

[5] Plant native to India

Dyes, especially acid, basic and reactive dyes, can escape conventional wastewater treatment due to their industrial purpose: resisting microbiological attacks and phenomena of photodegradation or chemical degradation [128]. In effluents from the textile, plastics, paint and ink industries, among others, dyes are found in colloidal, but non-sedimentable form, which makes their removal by mechanical processes difficult [104]. Depending on the acidic or basic nature of the dye and the ionic properties of the active agent, each coagulant is capable of removing a specific type of molecule. Using a modified tannin-based coagulant, Beltran-Heredia et al. [16] found high percentages of removal of anionic dyes of the azo and anthraquinonic classes, and did not obtain any effect on cationic dyes (methylene blue).

The removal efficiency of different dyes by the same coagulant is variable, even if they have the same structure. Szygula et al. [107] found different removal percentages in reactive dyes with the same amounts of sulphonic groups. The explanations for these differences are related to a) the acid-base properties of the dye (possible dissociations) or b) the aggregation mechanisms of the dye.

In the study of coagulant-dye systems, statistical analysis based on experimental planning is a tool for evaluating different conditions (pH, coagulant dosages, dye concentration, temperature, etc.) in order to determine the optimum operating conditions [63].

According to Table 10, the majority of studies are aimed at removing acidic dyes, for which natural coagulants generally show efficient decolorization. On the other hand, in some cases the results for basic dyes do not show good results [16,63] due to the repulsion of electrostatic forces between the dye and the coagulant in aqueous solution. However, other research [12] shows the possibility of removing this type of dye with acceptable effectiveness.

Basic dyes are considered the most toxic by ETAD (Ecological and Toxicological Association of Dyestuff Manufacturing Industry) [22]. In this study, basic dyes from the monoazo (Maxilon Blue 300 GRL), triarylmethane (malachite green) and thiazine (methylene blue) classes were chosen. To date, little research has focused on evaluating natural coagulants in cationic dyes [12]. Information on these dyes is presented in Table 11.

Previous research [12] has shown the ability of tannic acid and catechin, polyphenols present in various parts of plants, to induce discoloration and sedimentation of particles in aqueous solutions. The molecular interaction mechanisms proposed are ion-ion and ion-dipole relationships, taking advantage of the permanent dipoles resulting from the electronegativity of the oxygen in the functional groups or even the loss of hydrogen ions in the aqueous solution [94].

In this way, the evaluation of the decolorization of cationic dyes constitutes a broad field of research that can help solve a specific problem due to the toxicity and problems generated by the presence of these dyes in effluents. The following chapters present the methodology and results of the extraction processes carried out with the aerial part of cassava, and the decolorization tests conducted on some basic dyes.

Table 11. Main characteristics of the basic dyes tested

Category	Features	Problems generated in the effluent
Azo (mono)	• They account for 60% of existing commercial dyes (especially reactive dyes) and 43% of basic dyes. • The presence of sulphonated groups adds to its solubility in water • Biodegrade in anaerobic environments • It has a high standard of fixation and is highly resistant to light and humidity.	• Its lethal doses are less than 1.0 mg· 1(-) 1 • Their hydrolyzed forms produce carcinogenic amines • 15% of azo dyes are spilled without any treatment.
Triarylmethane	• They are intensely colored. • They are used as pH indicators • They are used as hydrological tracers • They account for 3% of industrially used dyes and 11% of basic dyes • They are mainly used in acrylic fibers due to their quick fixation	• Their molecular modifications in the effluent are potential carcinogens. • They cause metabolic alterations in fish present in contaminated waters. • Highly toxic to mammalian cells, promoting liver tumors
Taizina	• That's 5% of basic dyes • In addition to being used as dyes, they are also used as insecticides and in biological applications. • 60% of thiazinic dyes are basic	• Their lethal doses are less than 1.0 $mg{\cdot}l^{-}1$. • They are associated with nervous system problems and severe eye irritation

4 MATERIAL AND METHODS

Figure 15 shows a working scheme proposed for evaluating the coagulating activity of the aerial part of cassava. The diagram includes the processing steps commonly followed in research to extract protein or phenolic compounds from vegetables (Figure 13) [94]. The leaves and stem were studied separately in order to take advantage of the physico-chemical characteristics of each part studied.

4.1 Material

4.1.1. Cassava leaves and stems

Cassava (*Manihot esculenta* Crantz) leaves and stems were supplied by the Três Irmas Mill, located in the Três Barras community, in the Palhoça region, state of Santa Catarina. Both leaves and stems were collected from plants that were approximately four months old (time after planting). The fresh leaves were separated from the stem and transported in Styrofoam boxes directly to the laboratory for washing and drying, a process that was carried out up to 24 hours after collection.

The stems were previously dried and crushed in the mill. The crushed product was transported to the laboratory, packed in high-density polyethylene bags, sealed and stored at -20°C until it was used to prepare the crude extracts.

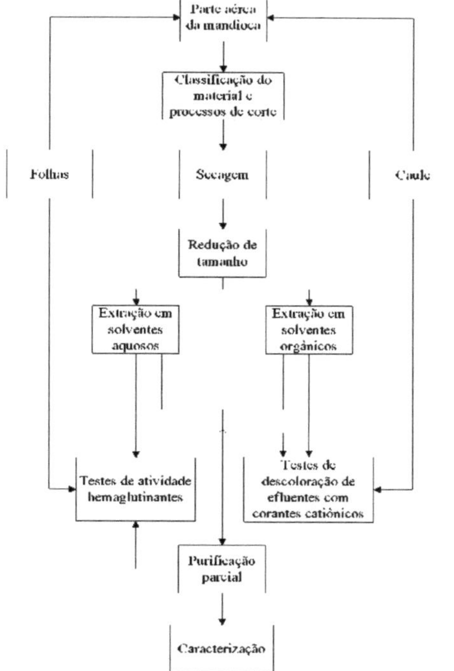

Figure 15. Proposed work scheme

4.1.2. Red blood cell samples

The accepted test for measuring hemagglutinating activity uses human, rabbit or rat blood. Since human blood is the most readily available, it was used as a standard. Bovine, chicken and pig bloods were used for possible application in the treatment of slaughterhouse effluents."

Type O+ human blood was collected from volunteers at the Santa Luzia Laboratory, Florianópolis, SC. Chicken blood was collected at the Sadia production plant in Chapecó, SC. The pig and bovine blood samples were obtained from the experimental slaughterhouse of the Instituto Federal Catarinense, Campus de Concórdia, Concórdia, SC. All samples were collected in 3.8% sodium citrate anticoagulant solution (1 ml of citrate solution per 9 ml of blood) to prevent clotting during transportation, which was carried out at refrigeration temperature (2-8°C).

In the laboratory, the mixtures were washed successively with buffered saline solution pH 7.4 in the proportion of 1 ml of blood to every 4 ml of saline solution and centrifuged at 1500× g at 20°C for 15 minutes. The supernatant (serum and hemolyzed material) was discarded and the precipitate washed. This procedure was repeated until a whole erythrocyte mass was obtained, free of serum or hemolyzed material. Finally, the volume of precipitate was diluted with buffered saline pH 7.4 until a 2% (v:v) erythrocyte suspension was obtained. The erythrocyte suspensions were refrigerated until they were used in HA tests [10,58].

4.1.3. Dyes

The dyes used for the studies were Methylene Blue, Malachite Green and Maxilon Blue GRL 300%, all of which are basic. Each of them has a different classification according to its Color Index (CI). Other characteristics of these dyes are presented in Table 12 and their structures are shown in Figure 16.

Table 12. Properties of the dyes used

Dye	CI code	Class	Molecular mass gmoΓ(1))	Maximum (g·absorbance (nm)
Methylene blue	52015	Thiazine	319,8	665
Blue Maxilon GRL 300%	11105	Monoazo	456,0	608
Malachite Green	42000	Triaryl methane	364,9	620

Source: [12,104,131,132]

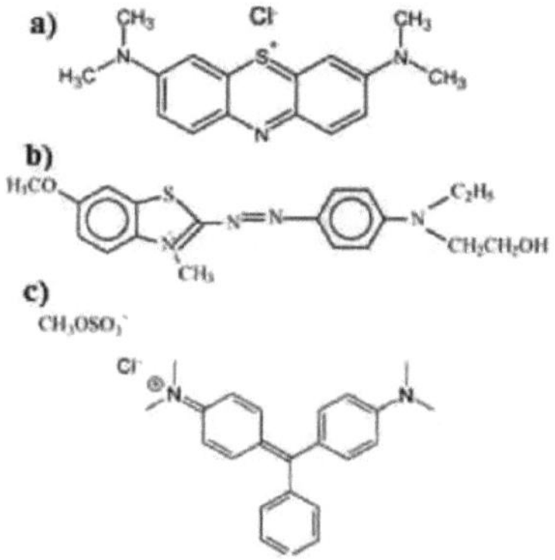

Figure 16. Molecular structures of the dyes a) Methylene Blue, b) Maxilon Blue GRL 300% and c) Malachite Green.

4.1.4. Other materials

Dialysis bags with an exclusion limit of 2 kDa and dimensions of 33*21 mm were supplied by INLAB. The bovine serum albumin (BSA), tannic acid and potassium hydrogenphthalate used as standards were SIGMA brand. The other reagents were of analytical purity from various brands.

4.2 Methodology

4.2.1. Primary processing

The fresh leaves brought to the laboratory were washed in 1% sodium hypochlorite and dried in the shade for 48 hours at room temperature (20-25°C). The stems were removed, as well as the petioles and other materials [30]. The leaves were then dried in an oven at 35-40°C until constant weight or until less than 10% moisture was obtained. The dried leaves were ground in an industrial blender until a flour with a maximum particle size of 1 mm was obtained. The flour was stored at refrigeration temperature (2-8°C) in sealed plastic bags or hermetically sealed glass jars [29].

Pieces of stems, leaves or any other type of material were removed from the raw stem sample. It was then dried in an oven at 35°C until constant weight or until it reached a maximum humidity of 10%. The dried material was ground in an industrial blender until a flour with a maximum particle size of 1 mm was obtained. The flour obtained was stored in the same way as the leaves (2-8°C).

4.2.2. Secondary processing

4.2.2.1. Extraction

The aqueous extraction was carried out at a ratio of 1:20 (g of ground leaves:ml of extracting agent) with sufficient mechanical agitation to suspend the solids in the solvent. In the case of the leaves, the extraction time and temperature conditions were optimized, based on previous studies [10,58]. Extraction times of between 15 and 120 min and temperatures of 10, 25 and 40°C were tested. As for the stem, since there were no previous studies on the extraction process, the temperature and extraction time conditions were established under which the protein and tannin yield obtained from the leaves was the best possible.

After extraction, the crude extract was filtered twice with Whatman N°.1 paper, resulting in a liquid free of suspended solids. If particulate matter remained in suspension, it was centrifuged (2500× g, 20 °C, 10 min) after filtration. The extracts obtained were stored in a refrigerator (2-8°C) and used for the different tests no more than 3 days after extraction.

The organic solvent extraction process was carried out for the two processed flours (leaves and stem). A ratio of 1:20 (g of flour:ml of extracting agent) was used, using 70% ethanol as the organic solvent [11]. The aqueous extract was double filtered using Whatman No. 1 filter paper followed by centrifugation under the conditions described above.

The crude extracts were concentrated [11,87] in a vacuum rotary evaporator at 60 °C, stirring speed: 40 rpm, vacuum pressure 60 mmHg, and distiller cooling to 10 °C. This procedure was carried out until 85% of the initial volume of the filtered extract had been reduced, taking approximately 45 minutes for every 450 ml of original extract.

The extracts were stored at 2 to 8°C and could be used up to a maximum of 3 days after extraction.

4.2.3. Ternary processing

In order to characterize the active principles present in the crude extracts, partial purification processes were carried out. They are described below:

4.2.3.1. Salt fractionation of crude leaf extract

Solid ammonium sulphate was added to the aqueous extract of cassava leaves until it reached 50% saturation according to Equation 1 [54].

$$g = \frac{533(S_2 - S_1)}{100 - 0{,}3S_2} \qquad (1)$$

Where:

G = Amount of ammonium sulphate per liter of solution

S1 = Initial percentage of ammonium sulphate saturation (0%) S2= Final percentage of ammonium sulphate saturation (50%) The value of 533 corresponds to the solubility of ammonium sulphate at 20 °C (533 $g \cdot L^{(-1)}$). The values of 100 and 0.3 were determined from the fact that a saturated solution of ammonium sulphate requires 761 $g \cdot L^{(-1)}$.

The crude extract was kept in an ice bath with magnetic stirring during the addition of the sulfate, which was done slowly to eliminate the possibility of denaturation due to surface tension phenomena. Stirring was maintained for 30 minutes after the last addition of salt to ensure equilibrium [54]. The mixture was then left to rest at 2-8°C overnight. The extract was then centrifuged (8000× g, for 15 min at 10°C) [10]. The precipitate was redissolved in phosphate buffer solution pH 7.4 and dialyzed exhaustively against the same solution, changing twice a day for 48 hours, stirring constantly.

In the precipitation process, ammonium sulphate was first added slowly with constant stirring to avoid the formation of bubbles, which indicates the denaturation of proteins at the air-water interface; secondly, a specific amount of sulphate was added, taking into account the initial and final density of the solution, in order to maintain the desired percentage of saturation [133].

Part of the dialyzed suspension was subjected to hemagglutination tests, and the other was dried at 35°C and stored in a refrigerator until use in HA tests or FTIR characterization tests. For HA tests, the dialyzed sample was suspended in as little water or buffer solution as possible.

4.2.4. Characterization

4.2.4.1. Humidity

Moisture was determined in triplicate using the method of water loss by drying to constant weight at 105°C of the fresh leaves, the previously dried stem and the flours obtained after grinding [134].

4.2.4.2. Protein content

The protein content of the extracts was determined by the Bradford dye-protein binding method [135] using Bovine Serum Albumin (BSA) as a standard (ANNEX 1).

4.2.4.3. Tannin content

The determination of tannins was carried out according to the method summarized by Barman et al. [136], which includes a previous removal of pigments from the solid samples with diethyl ether and acetic acid. The measurement of phenolic compounds was carried out using 70% ethanol (to simulate the real conditions of the extract) and acetone (to ensure the greatest possible removal) as solvents. The tannin content is determined by the difference between total phenolics, measured using the Folin-Ciocalteu method (Makkar et al. [92]), and non-tannic compounds, determined using polyvinylpyrrolidone (PVP) as a binder. The standard curve was drawn up using tannic acid (ANNEX 1). The values were reported as tannic acid equivalents (dry basis). Condensed tannins were analyzed using the Butanol-HCl method according to Porter et al [137]. The results were calculated using Equation 2 and reported as leucocyanidin equivalents [136]:

$$\% \text{ TC} = \frac{A_{550\text{ nm}} * 78{,}26 * F}{\%_{MS}} \qquad (2)$$

Where:

%TC = Percentage of condensed tannins.

A = Absorbance of the sample at 550 nm.

F = Dilution factor in the spectrophotometric measurement.

$\%_{DM}$ = Percentage of dry matter in the sample.

Hydrolysable tannins were calculated as the difference between total tannins and condensed tannins.

4.2.4.4. Determining the zero charge point

The ethanolic extract, previously concentrated and frozen (-20°C) for 24 hours, was lyophilized. The solid was stored in an inert atmosphere until the FTIR spectrum was carried out and the zero charge point determined.

The determination was made according to Cerovic et al [138]. Samples of 0.200 g of the lyophilized solid obtained from the ethanolic extract of the stem were placed in flasks with 20 ml of 0.1 N NaCl solution, previously adjusted to different pH values by adding 0.1 M HCl or NaOH. The hermetically sealed flasks were kept under constant agitation (90 rpm) in a Dist DI-940 shaker at 25°C. The pH was measured using a Digimex DM-23 pH meter.

The pH was monitored every 1 hour until constant values were obtained, a point reached after about 4 hours. For convenience, stirring was maintained for 24 hours to ensure equilibrium of the solutions. The final pH values were plotted as a function of the initial pH values. The intersection of the curve with the y=x line indicates the point of zero charge of the purified solid.

4.2.4.5. FTIR spectra

In order to obtain structural information on the extracted compounds, FTIR analysis was carried out on the following samples:

- Aqueous extract of the leaves, precipitated with ammonium sulphate, dialyzed and dried.
- Ethanolic extract of the stem, concentrated and freeze-dried.

Readings were taken on KBr pellets at the Analysis Center of the UFSC Chemistry Department on a System 2000 spectrophotometer, PERKIM & ELMER 16 PC.

4.2.5. Hemagglutinating activity

The determination of HA was carried out using different erythrocyte suspensions prepared according to the methodology described by Silva [10]. Microtiter plates, containing 4 rows of 6 wells each, were filled with 100 µl of phosphate buffer solution pH 7.4 and then added with an equal volume of extract sample in the first wells of the row. The sample was subjected to serial dilution, base 2 in triplicate, with homogenization and transfer of 100 µl to the next well up to the penultimate well in the row. In this way, the last well contained only saline solution and was used as a blank.

Each well was then incubated with 100 µl of the 2% erythrocyte suspension prepared earlier. Agglutination was observed macroscopically after incubating the samples for 30 min at 37°C [139] with a confirmed reading 12 hours after storage at 4°C [38]. The results were expressed as specific HA (SHA), which is the ratio between the HA and the protein content of the extract. The calculation was made from the inverse value of the titre of the highest dilution that still had HA and the protein content measured by the Bradford method [135]. Below is an example of the calculation:

- Maximum dilution with visible agglutination: 2
- Dilution title: 2^2=4

- Volume of sample used: 100 µl
- Protein content of the sample: 10.3µg· 100 μl^{-1}

Calculation of AHE:

$$AHE = \frac{4UH}{100\mu l_{EXTRATO}} * \frac{100\mu l_{EXTRATO}}{10,3\mu g_{PROTEINA}} * \frac{1000\mu g_{PROTEINA}}{1g_{PROTEINA}} = 388,3\, {}^{UH}/_{g_{PROTEINA}} \quad (3)$$

In order to optimize the extraction time and temperature conditions, extractions were carried out in triplicate with extracting agent volumes of 1 liter. Aliquots of crude extract of 5 ml were taken every 15 min, filtered and centrifuged (2500× g, 20 °C, 10 min) in which the percentage of protein extracted and the HA were evaluated. The conditions under which the HA was highest were used for subsequent extractions, both on the leaves and on the stem. The extracts with the highest HA were stored at 10, 25 and 40°C. The AH was monitored every 12 hours for 7 days, and the time when the AHE reached half of the initial value was determined (half-life time).

4.2.6. Dye removal tests

Standard solutions of 1000 mg· $^{L(-1)}$ of the dyes were prepared with distilled water, and dilutions of 10, 40 and 100 mg· $^{L(-1)}$) were used for testing. The pH was then adjusted by adding NaOH or 0.1 M HCl.

4.2.6.1. Discoloration tests

Four types of extract, aqueous and ethanolic, both leaf and stem, were preliminarily tested in triplicate in test tubes containing 10 ml of dye solution. The extract with the best visual result was used for the subsequent decolorization study in a jar-test (Quimis).

The best pH conditions, initial dye concentration, salt concentration (NaCl), stirring time for the fast and slow mixing steps and dosage of crude extract were determined using methylene blue in 200 mL jar-test volumes. The optimization of the system with these variables used a working temperature of 20°C and stirring speeds of 170 and 50 rpm for the fast and slow steps, respectively.

Figure 17 shows the layout of the Jar-tests; the addition of salt was done before the final pH adjustment due to the strong influence of NaCl on the solutions studied. The extract was added during the rapid mixing stage to ensure homogeneity in the dye solution. After rapid mixing, the solutions were left to stand for 1 hour, and discoloration was monitored at 1, 24 and 48 hours by keeping the dye solutions with the extract at complete rest.

The dyes Maxilon Blue GRL 300% and Malachite Green were also tested, varying only the pH and extract dosage. The other variables were set according to the optimization made with methylene blue. The effect of adding an alkalizing agent (Na2CO3) to the three dyes was also tested. This agent, as well as the extract, was added during the rapid mixing stage.

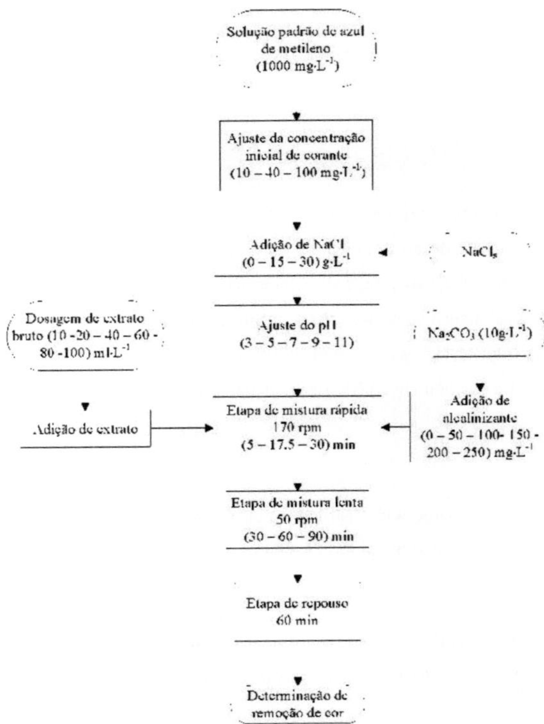

Figure 17. Jar-Test procedure for methylene blue removal

After determining the best conditions for each dye, the color, COD and toxicity parameters of the solutions were determined in triplicate.

4.2.6.2. Determining the percentage of color removal

Color was determined using a spectrophotometer (Bel Photonics SP 1105) with readings at the wavelength of maximum absorbance in the visible range for each dye according to Table 12. The dye calibration curves relate the concentration of the dye in the aqueous solution ($mg\cdot ^{L(-1)}$) to the absorbance (ANNEX 1); based on these curves, the percentage removed from the initial solutions was calculated. For both the standard curves and the removal tests, the necessary dilutions were made to satisfy the Lambert-Beer Law, always working in an absorbance range in which linearity was guaranteed.

4.2.6.3. Determination of COD

The increase in organic matter generated by the addition of the extracts to the effluent was determined by evaluating the chemical oxygen demand (COD). The determination was made under optimum working conditions.

The procedure followed was that corresponding to the closed reflux colorimetric method [140], which has as its basis the oxidation of the dichromate ion during the digestion of the samples at 150°C for two hours. The change from hexavalent chromium (VI) to trivalent chromium (III), with absorbance at 600 nm, makes it possible to measure the intensity of the color and thus the COD of the sample. The color was measured using a spectrophotometer (Bel Photonics SP 1105). Potassium hydrogen phthalate was used to make the calibration curve (ANNEX 1), knowing that hydrogen phthalate has a theoretical COD of 1,176 mg $O2 \cdot mg^{-1}$.

4.2.6.4. Toxicity tests

The toxicity of the raw and treated dye solutions was assessed using the *Artemia salina* test (method adapted from Pérez and Lazo [141]): Multi-well plates were filled with 5 ml of dye solution; subsequently, 10 larvae of the microcrustacean were added to each well (taking care that the volume of water from which the microorganisms were taken was no greater than 0.1 ml). After 24 hours of incubation at 25°C, the number of dead larvae in each condition was counted. Analyses were carried out in quadruplicate.

In addition, different concentrations of the extract chosen for decolorization were evaluated in water in order to determine the CL50, which is the concentration of extract at which 50% of the initial larvae are found dead. The procedure was identical to that used for the dye solutions.

For the hatching and growth of the microcrustaceans, Furlan's instructions were followed [22]: 1 g of *Artemia salina* eggs was added to 0.5 liters of water previously mixed with 16 g of sea salt. The solution was incubated for 60 hours at 30°C with light and air.

4.2.7. Statistical data analysis

When optimizing the extraction conditions (time and temperature), the results for protein content and HA were statistically evaluated using the Tukey test and analysis of variance (ANOVA) for first-order (non-interactive) effects.

Fractional factorial design $2^{(k-p)}$ with fraction 1/2 for 6 factors (k=6, p=1) was used in the methylene blue decolorization tests. This type of design makes it possible to separate and estimate both the main effects and the interactions between the factors studied. In addition, the factors were evaluated in triplicate at the central point. The use of the central point makes it possible to evaluate the significance of the effects both in screening designs (complete or fractional) and in response surface methodologies [142], the latter being presented for the second-order relationships of greatest importance according to the results.

Generally speaking, the results shown in the various graphs and tables in this work present mean values and standard deviations for the tests, most of which were carried out in triplicate.

5 RESULTS AND DISCUSSION

5.1 Humidity

Table 13 shows the moisture content of the materials in their initial condition (fresh leaves and stem previously dried in the mill) and after primary processing (leaf and stem flours).

Table 13. Moisture content of cassava leaf and stem samples

Sample	Humidity (%)[1]
Fresh leaves	63,0±1,0
Pre-dried leaves	9,0±1,7
Leaf meal	11,3±1,8
Previously dried stem	11,9±0,4
Stem flour	13,7±0,1

[1] Tests carried out in triplicate

The results, on a dry basis, shown in the following sections, correspond to percentages or contents calculated with respect to the moisture content of the leaf and stem flours.

5.2 FTIR spectra

The FTIR spectrum of the solid obtained after concentrating and freeze-drying the ethanolic extract of the cassava stem is shown in Figure 18. The broad band in the range 3600-3100 cm^{-1} corresponds to OH groups, which are attributed to phenolic groups or hydrogens from water molecules attached to these groups. In addition, the presence of residual alcohols can also generate vibrations in this range. The presence of phenolic groups is confirmed by absorption in the 1440 to 1220 cm^{-1} and 1260 to 1000 cm^{-1} regions [143].

The small peaks in the 2950-2850 cm^{-1} region are associated with methylene groups (-CH2-). In addition, vibrations from C-H groups present in aromatic rings also generate absorption in this region. Vibrations near 1600 cm^{-1} are characteristic of -C=C- bonds. The

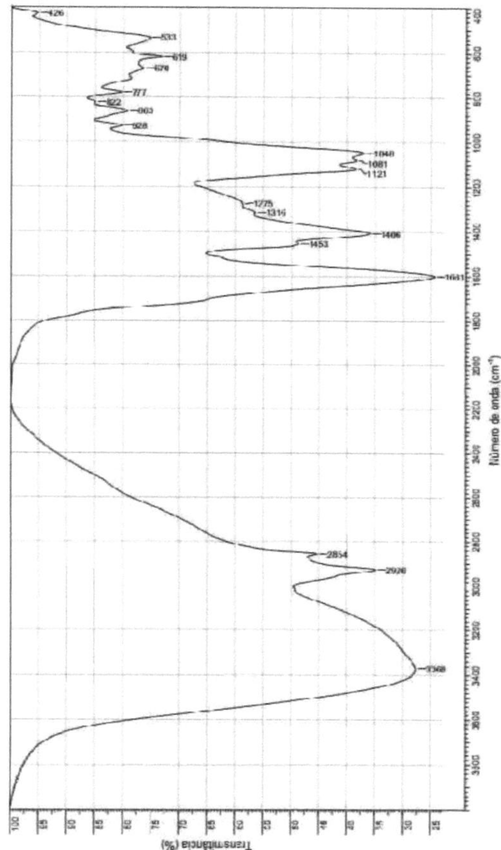

Figure 18. FTIR spectrum of the concentrated and lyophilized cassava stem ethanolic extract

Characteristic vibrations of the C-C bonds of phenolic groups are observed in the 1500-1400 cm^{-1} region. The peaks in the 1160975 cm^{-1} region are due to asymmetric C-O-C group shifts and C-H deformations. Vibrations due to deformations of the C-H bonds in aromatic rings generate absorption in the 835-650 cm' range [1][143].

Figure 19 shows the spectrum of the aqueous extract obtained from cassava leaves after precipitation with ammonium sulphate, dialysis and drying at a controlled temperature (35°C). The presence of a wide range of vibrations near 3400 cm^{-1} suggests the presence of O-H groups from phenolic compounds or water remaining in the solid evaluated, the intensity of this region is much lower than in the spectrum of the ethanolic extract (Figure 18). The smaller amount of polyphenols and tannins removed with the aqueous solvent may explain the decrease in vibrations in the range around 3400 cm^{-1}. The slight absorption at 2932 cm^{-1} and the peak at 1415 cm^{-1} are related to the existence of -CH2- groups [143].

The existence of a vibration at 1640 cm^{-1} suggests several alternatives for analyzing the spectrum in Figure 19, such as the presence of primary amides, where this vibration is mainly due to C=O groups and angular deformations of N-H groups. Lectins from other sources have been explained by the presence of vibrations in the range described above. The exact value of the frequency is determined by the molecular geometry of the polypeptide [144].

The set of vibrations at 3246 cm^{-1} (-NH2), 1640 $cm^{(-1)}$ (N-H), 1415 $cm^{(-1)}$ (-CH2-) and 1095 cm(-1) (C-N) may be related to the presence of amines. The two peaks in the 3200 to 3500 cm^{-1} region are characteristic of primary amines due to the double hydrogen bond. Finally, the absorption at 1640 cm^{-1} together with the presence of peaks to the left of 3000 $cm^{(-1)}$ (3060 cm^{-1} in the spectrum) may also represent the presence of C=C double bonds [143].

Analysis of the spectra shows a greater presence of aromatic and phenolic compounds in the ethanolic extract (sample analyzed from the stem), and the possibility of nitrogen compounds in the solid obtained from precipitating the aqueous extract of cassava leaves with ammonium sulphate, suggesting the presence of protein compounds. However, these hypotheses need to be confirmed using more precise characterization techniques, such as chromatographic analysis.

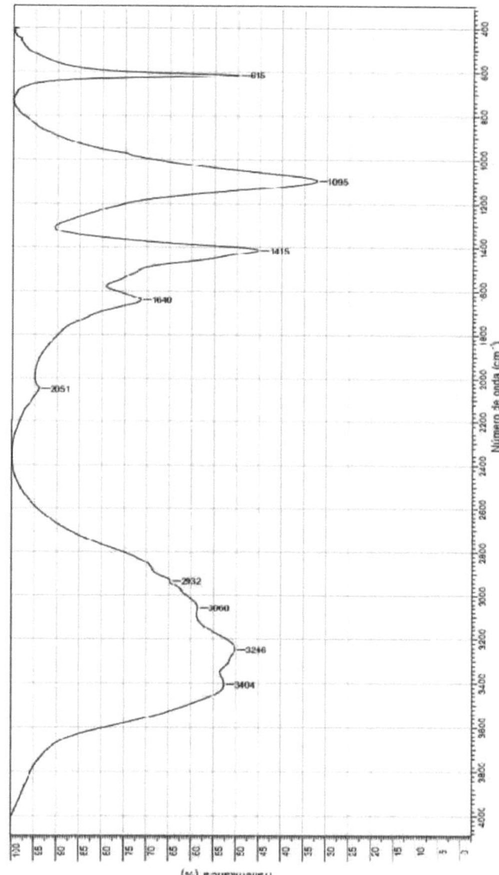

Figure 19. FTIR spectrum of aqueous cassava leaf extract precipitated with ammonium sulphate, dialyzed and dried.

As can be seen in the spectra, there are various functional groups (phenolic, carboxylic, alcoholic, aromatic rings, nitrogen compounds, etc.), many of which are directly related to the possibility of agglomerating substances by various mechanisms (adsorption, neutralization of charges, formation of bridges, etc.) and can be a starting point for describing coagulation and flocculation processes [106].

5.3 Protein and tannin content

Once the extraction conditions had been defined, the protein and tannin contents of the stems and leaves were determined. Table 14 summarizes these results. The standard curves for BSA in the case of protein analysis and tannic acid in the case of phenolic and tannin analysis can be found in ANNEX 1.

Table 14. Protein and tannin content of extracts[1]

Determination	Leaves		Stem	
	Aqueous extract	Ethanolic extract	Aqueous extract	Ethanolic extract
Proteins ($mg{\cdot}g^{-1}$)[2]	2,62±0,82	8,83±2,33	1,43±0,23	2,60±0,48
Phenolics total ($mg{\cdot}g^{-1}$)[2]	7,35±0,86	21,22±3,78	1,29±0,36	2,01±0,64
Total tannins ($mg{\cdot}g^{-1}$)[2]	4,05±0,65	13,69±0,10	0,17±0,05	0,93±0,23
Hydrolysable tannins (%)[3]	0,79±0,05	13,15±0,89	0,91±0,15	9,01±1,45
Condensed tannins (%)[3]	99,21±0,05	86,85±0,89	99,09±0,15	90,98±1,45

1Extraction conditions: time 60 min, temperature 25°C, solid:solvent ratio:

1g:20 ml

[2]mg of protein-g of solids^{-1} (Results on a dry basis for ground raw materials)

[3]Percentages with respect to total tannins

In general, extraction with organic solvents produced better yields of protein and phenolic compounds. The leaf meal contains higher amounts of protein and phenolic compounds than the stem. Although there are no specific studies on the latter, nutritional supplement preparations containing mixtures with different proportions of stem decrease their protein content as the percentage of stem increases [46].

In the case of phenolic compounds, the levels obtained are in the range of results found in the literature [3,34], which were obtained using methanol as a solvent [9]. Table 15 shows the results obtained using 70% acetone as the extracting agent, which was considered a reference, comparing the extractions made with water and 70% ethanol.

For tannins (Table 14), the values determined in the leaves are also within the range of the results presented in the literature (240 mg· g(-)1 of leaves) [145,146], where the differences found can be attributed to the degree of ripeness at the time of harvest [9].

The significant amount of condensed tannins in the leaves is a common characteristic of cassava [8]. It can be seen from the results presented above that a high percentage of the phenolic compounds present in the aerial part of cassava correspond to tannin molecules, a common characteristic of this plant's Iytochemical profile [76].

Table 15. Recovery of polyphenols and tannins for different solvents[1]

Solvent	Polyphenols (mg· g(-1))[2]		Tannins (mg· g(-1))[2]	
	Content (mg· g(-1))2	Recovery (%)	Content (mg· g(-) 1)2	Recovery (%)
Leaves				
Water	7,35±0,86	20,5	4,05	15,4
Ethanol 70%	21,2±3,78	59,3	13,69	52,1
Acetone 70 %	35,8±2,91	100,0	26,26	100,0
Stem				
Water	1,29±0,36	23,7	0,17	5,2
Ethanol 70%	2,01±0,64	36,7	0,93	28,9
Acetone 70 %	5,47±1,45	100,0	3,23	100,0

[1]Extraction conditions: 60 min, 25°C, solvent-solid ratio: 1g:20 ml

[2]The results are expressed on a dry basis, based on the solvent-solid ratio.

The results of the protein and polyphenol quantification confirm the information obtained from the FTIR spectra.

5.4 Hemagglutinating activity

Figure 20 shows the protein contents obtained; extractions made at 40°C showed higher protein contents. Two points should be taken into account to explain this: the plant cell is weakened by temperature, decreasing its selective capacity, and the solubility of other non-binding proteins may have increased with the rise in temperature.

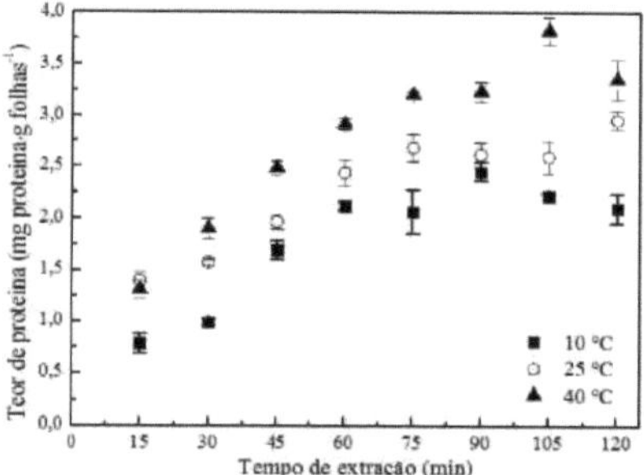

Figure 20. Protein content in aqueous extracts of fresh cassava leaves at different temperatures and extraction times.

The statistical analysis (Table 16) indicates a significant difference in extraction temperature, suggesting an important influence of this variable on the final protein percentage.

As for the extraction time, there is a first period (15 to 75 minutes) where there is a progressive increase in protein content, with the increase being more easily noticeable from 15 to 60 min. From 60 to 75 min, the increase in content is lighter and the results obtained do not differ statistically. For extractions carried out at 90 min or more, the percentages of protein extracted did not vary significantly, indicating that stabilization had been achieved. Similar values were found in previous studies [58]. The maximum protein contents were reached in the 90-105 min range for the different temperatures tested.

Table 16. Protein content (mg protein· g leaves^{-1}) for different temperatures and extraction times[1]

	Extraction temperature (°C)[2]		
Extraction time (min)	**10**	**25**	**40**
15	0.79±0.10A^{a}	1.40±0.09B^{a}	1.31±0.09C^{a}
30	0.99±0.03A^{b}	1.58±0.03B^{b}	1.90±0.10C^{b}
45	1.69±0.09A^{c}	1.97±0.07B^{c}	2.48±0.07C^{c}
60	2.12±0.04 A^{d}	2.43±0.13B^{d}	2.92±0.06C^{d}
75	2.06±0.21A^{de}	2.69±0.14B^{de}	3.21±0.09C^{de}
90	2.45±0.09A^{de}	2.62±0.09B^{e}	3.24±0.09C^{e}
105	2.22±0.09A^{e}	2.59±0.09B^{e}	3.82±0.09C^{e}
120	2.10±0.09A^{e}	2.97±0.09B^{e}	3.36±0.09C^{e}

[1]Values on a dry basis for cassava leaf meal [2]Values with the same upper and lower case letters in a row or

column respectively have no significant difference according to Tukey's test (p≤ 0.05).

In order to assess the selectivity of the lectin, the aqueous extract of cassava leaves was tested on various types of erythrocytes. HA was detected in suspensions of human, bovine and chicken erythrocytes (Figure 21 and Figure 22). It can be seen in Figure 22 that increasing the extraction temperature from 10 to 25°C leads to an increase in AHE for human erythrocytes. Extractions carried out at 40°C showed a decrease in activity when human (Figure 21) and bovine erythrocytes were tested; this behavior may be associated with the loss of stability of protein compounds [39], including lectins, whose highest content was obtained at 40°C (Figure 20). No activity was detected in the pig blood samples. Several studies in the literature indicate the selectivity of lectin compounds, taking advantage of these characteristics to use them in medical applications as tracers [43,44]. In this way, the variations found in AHE, as well as the total absence in some cases, suggest that the lectin present in the aerial part of cassava is specific in terms of the molecules to which it binds. The large standard deviations seen in Figures 21 and 22 are related to the method used to calculate the HA, which is based on the titre of the highest dilution with activity. Thus, variations from one dilution to another in triplicate tests can cause this type of deviation.

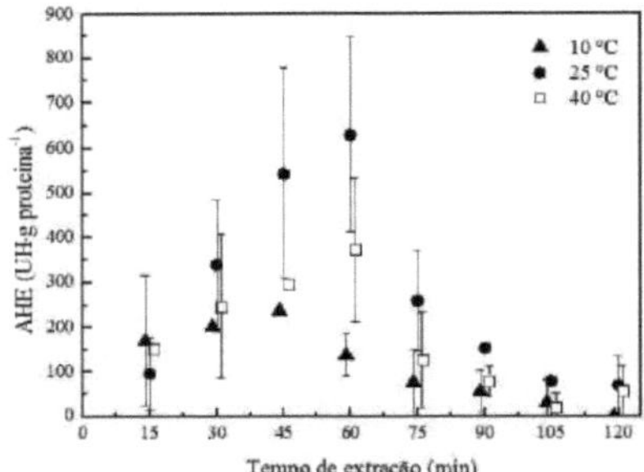

Figure 21. AHE of the aqueous leaf extract on human erythrocytes for different extraction temperatures.

According to Figure 22, increasing the extraction time beyond 60 min promotes a decrease in AHE, so that in extractions longer than 90 min in some cases no activity is detected. This loss of AHE can be explained by possible bonds between the protein and sugars present in the extract, since the extraction process is not selective [37]. According to Chakrabarti and Podder [147], lectin-sugar bonds have a low reaction rate and are controlled by diffusion phenomena and the formation of multiple bonds. Denaturation of the lectin can be ruled out by analogy; previous research has reported denaturation phenomena at temperatures above 70°C [148].

Compared to Figure 20, it is important to note that the AHE peak is reached before the time at which the protein content is maximum, and there is a possibility that other types of non-binding proteins are being extracted at long extraction times. Previous studies indicate that lectins make up 2-10% of soluble proteins and their concentration is directly related to the intensity of AHE [38].

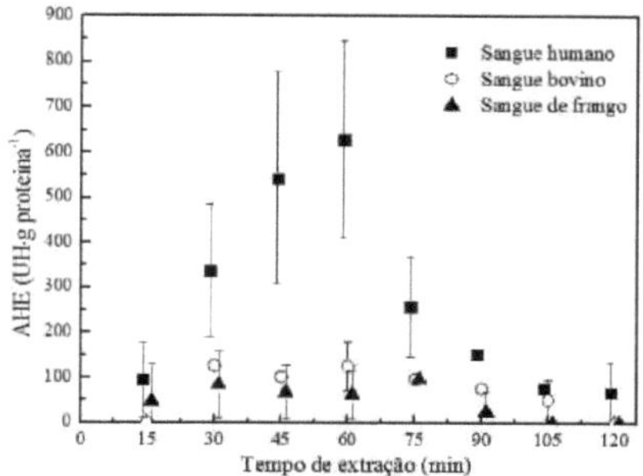

Figure 22. AHE of the aqueous leaf extract as a function of time for the types of erythrocytes in which AH activity was detected. Extraction conditions: 25°C.

According to the statistical analysis, there were no significant differences in extraction time for the temperatures of 10 and 40°C. In relation to extractions carried out at 25°C, an influence of extraction time was observed for human and bovine erythrocytes, especially where the highest AHE values are obtained. For chicken erythrocytes, no dependence on time was detected. Table 17 shows the average AHE values for extractions carried out at 25°C; it can be concluded that extraction times of between 30 and 75 minutes guarantee the highest activities.

By performing a similar analysis to compare the extraction temperatures, it can be said that in the case of chicken erythrocytes there were no significant differences in the AHE obtained. In the case of human and bovine erythrocytes, differences were found at extraction times of 30 and 75 minutes; moreover, this time range is where the highest activities are found, suggesting that the greater the lectin's ability to agglutinate, the more noticeable the difference when using one or other temperature for extraction, with the process being more efficient at 25°C (Figure 21).

Table 17. Significant differences according to Tukey's test for the extraction time of the aqueous leaf extract (Extraction temperature: 25°C).

Time of extraction (min)	AHE(UH· g protein^{-1}) and erythrocyte type[1]		
	Human	Cattle	Chicken
15	95,2 ± 82,5^{a}	N,D	47,6 ± 12,5^{a}
30	338.5 ± 146.6^{ab}	126,9 ± 0,0^{b}	84,6 ± 33,3^{a}
45	542.8 ± 135^{bc}	101.8 ± 0.0^{bc}	67,9 ± 48,8^{a}
60	628.7 ± 217.8^{bd}	125,7 ± 54,4^{b}	62,9 ± 44,4^{a}
75	258.6 ± 112^{acd}	97 ± 0.0^{bd}	97 ± 0,0^{a}
90	152,8 ± 0,0^{a}	76.4 ± 0.0^{be}	25,5 ± 4,1^{a}
105	77,1 ± 0,0^{a}	51.4 ± 24.5^{acde}	N.D
120	67,4 ± 37,4^{a}	N.D[2]	N.D

[1]Values with equal letters in a column have no significant difference according to Tukey's test ($p \leq 0.05$)

[2]N.D. = AH not detected

Table 18 summarizes the highest AHE found for the different types of erythrocytes tested. These results were used to establish the extraction conditions prior to partial precipitation with ammonium sulphate and to optimize the temperature conditions for storing the extracts.

Table 19 shows the effect of isoelectric precipitation on AHE. After the addition of ammonium sulphate and subsequent dialysis, there was a general increase in the HA of the human and bovine erythrocyte samples. The purification process concentrates the active ingredient, resulting in a higher final HA.

Previous studies have linked protein recovery to the activity achieved. Silva [30] found a protein recovery of between 50 and 60% in the sulphate precipitation stage and found AHE values of 2800 UH· $^{g(-1)}$ of protein; meanwhile Pereira [58], with recoveries of around 20%, obtained activities in the range of 8001000 UH· $^{g(-1)}$ of protein. In the present study, the average recovery percentage was 60%. Compared to the studies mentioned above, the activities found were in the range of 200900 UH· g of protein^{-1} (Table 19). One explanation for the decrease in AHE may be the use of a dialysis membrane with a low exclusion limit (2 kDa), which can retain proteins other than lectins [58].

Table 18. Maximum AHE of the aqueous leaf extract for the different types of erythrocytes.

Type of erythrocyte	Extraction temperature (°C)	Maximum AHE (UH- g protein^{-1})	Extraction time (min)
Human	10	258,6	45
	25	628,7[2]	60
	40	152,8	60
Cattle	10	97,0	60
	25	125,7[2]	60
	40	76,4	60
Pork	10	N.D[1]	--
	25	N.D	--
	40	N.D	--
Chicken	10	67,2	30
	25	97,0[2]	75
	40	N.D	--

[1]HA not detected

[2]Maximum specific hemagglutinating activity (AHE_o).

Table 19. AHE (UH· g protein^{-1}) of the crude leaf extract precipitated with ammonium sulphate and subsequently dialyzed.

Extract	Type of erythrocyte			
	Human	Cattle	Chicken	Pork
Gross	628,7± 217,8	125,7±54,4	N.D	N.D
Purified in distilled water	912,4± 430,1	608,3 ± 0,0	N.D	N.D
Purified in phosphate lid	825,5 ± 0,0	619,1 ± 291,8	N.D	N.D
Purified and dried in lid phosphate[1]	517,8 ± 244,1	258,9 ± 122,0	N.D	N.D

1After dialysis, the sample was dried at 40°C for 3 hours and redissolved for the HA test.

The extracts with the highest activity (Table 18) were used to determine the best storage temperature conditions and, in parallel, the time when AHEo decreased to half its initial value. Table 20 shows the results obtained for different storage temperatures. Lower temperatures favored the stability of the lectin, denoted by the longer half-life. This is probably due to the inhibition of phenomena commonly associated with physical and chemical degradation and precipitation of compounds. Subsequent purification processes of the lectin present would increase the half-life times in relation to AHE. As an example, Figure 23 shows the behavior of AHE with storage time for human erythrocytes.

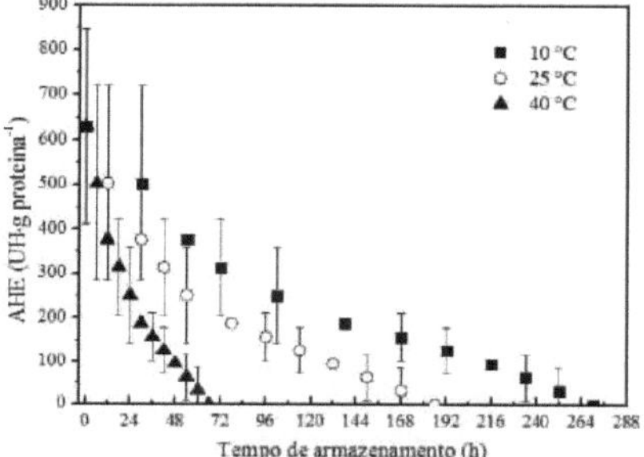

Figure 23. Stability of AHEo in human erythrocytes at different storage temperatures.

Table 20. AHE half-life for the different types of erythrocytes tested.

Type of erythrocyte	Storage temperature (°C)	AHEo[1] (UH-g protein^{-1})	AHE half-life (h)
	10	628,7	72
Human	25	628,7	42
	40	628,7	18
	10	125,7	138
Cattle	25	125,7	48
	40	125,7	18
	10	97,0	87
Chicken	25	97,0	60
	40	97,0	30

[1]Maximum specific hemagglutinating activity.

The behavior for bovine and porcine erythrocytes was similar, with greater stability at 10°C, and the extracts used in hemagglutination and discoloration tests had a maximum of 3 days of storage, considering the longest half-life at 10°C for human erythrocytes (72 h) as a parameter.

Based on the behavior of the hemagglutination phenomenon, the conditions for the ethanolic and aqueous extractions of both the leaves and the stem were defined; the working temperature was set at 25°C and the extraction time at 1 hour.

5.5 Discoloration tests

Table 21 shows the results of the preliminary tests carried out on AM bleaching.

Table 21. Color removal in preliminary tests with AM (%)

Extract[1]	Initial dye concentration (mg· L(-1))2

	10	40
El - Aqueous leaf extract	14,7	4,1
E2 - Ethanolic extract of leaves	20,4	12,9
E3 - Aqueous stem extract	7,3	4,0
E4 - Stem ethanolic extract	11,5	8,2
E5 - Concentrated aqueous phase obtained from the ethanolic extract of the stem	39,2	32,0

[1] Extract dosage: 20 ml extract-l dye^{-1}.

[2]Initial pH of 6.1 and 5.6 for the 10 and 40 mg-L^{-1} solutions respectively.

[3] Not determined.

From these results, it can be seen that the aqueous phase obtained from the concentration of the ethanolic extract of the stem (E5) was the one that showed the best results. The concentration of the extract and of the condensed active ingredients may be related to the coagulant activity shown in these preliminary tests.

Figure 24 shows the result of using extract E5 in AM solutions of different initial concentrations. Discoloration of the treated samples is observed. Along with the gradual discoloration, increasing the incubation time leads to the formation of precipitated flakes. This indicates that the extract induces the coagulation-flocculation phenomenon, which relies on initial molecular interactions between the extract and the dye in aqueous solution.

Figure 24. Coagulation and discoloration of AM solutions. A) 40 mg$\cdot^{L(-)}$ 1; B) 10 mg$\cdot^{L(-1)}$. Extract dosage 20 ml$\cdot^{L(-1)}$ of dye solution.

Photographs taken after 48 hours of incubation at 25°C.

The formation of sedimented flocs in similar decolorization processes is also related to the presence of phenolic compounds with low pKa values. Studies using synthetic compounds rich in tannins (tannic acid and catechin) confirm their ability to aggregate organic pollutants of a cationic nature [12]. In the results obtained, the sedimentation capacity of the flocs formed in the coagulation stage is slow and, for this reason, it is industrially recommended to use a flocculating agent after the coagulant to speed up the process of decanting the agglomerated solid [98].

The presence of slight yellow colors observed in the crude extracts is due to pigments extracted together with the active ingredients, which interfere with the discoloration studies. However, the influence on the color measurement and appearance of the treated solutions was not significant. This type of problem has been reported in other studies [105].

In the case of the ethanolic extract of the leaves (E2), even though it had the highest levels of protein and tannins, the extraction of various pigments related to the alcoholic phase of the extractor interfered with the measurement and made this extract unusable for removing color.

Although some research suggests that tannins are cationic due to the presence of tertiary amines [7], the removal capacity of the E5 extract may also be related to the presence of polyphenols with an amphoteric nature, such as phenolic groups; in addition, concentrated extracts have an increased availability of phenolic groups, which improves efficiency in the coagulation process [94].

5.5.1. Methylene blue (AM)

With the best condition obtained in the previous study (concentrated aqueous phase of the ethanolic extract of cassava stems, E5), Jar-Test trials were carried out according to a 2^{6-1} experimental design (trials 1-32 in random order) with a central point (trials 3335), which are shown in Table 22. Optimizing the factors studied will allow the best decolorization conditions to be obtained. The first and second order effects (interactions between two first order effects) were evaluated in the statistical analysis.

The traditional one-factor-at-a-time method proves difficult when there are several factors that may or may not interact with each other. Even when traditional analysis is used to determine the predominant factors, it is difficult to find the optimum values when the interactions between the factors are taken into account. To solve this problem, experimental design offers a better alternative for studying the effects of variables and their responses with a smaller number of experiments [16].

Table 22. Experimental design and results of the AM decolorization study.

Essay	**Factor**						
	1- [AM] (mg· L(-)1)	**2- pH**	**3 -Extract dosage (ml· L(-) 1)**	**4 -[NaCl] (g· L(-1))**	**5 -TAR[1] (min)**	**6 -TAL[2] (min)**	**% of removal**
7	10	9	40	0	5	30	40,7±2,9
16	40	9	40	30	5	30	18,3±2,7
27	10	9	10	30	30	90	12,9±0,3
12	40	9	10	30	5	90	0,0±0,0
32	40	9	40	30	30	90	38,4±3,0
31	10	9	40	30	30	30	34,9±3,4
17	10	5	10	0	30	90	6,8±1,2
30	40	5	40	30	30	30	21,3±1,4
20	40	9	10	0	30	90	6,7±1,8
29	10	5	40	30	30	90	27,4±1,7
28	40	9	10	30	30	30	10,5±0,6
19	10	9	10	0	30	30	33,3±3,4
22	40	5	40	0	30	90	2,7±0,1
10	40	5	10	30	5	30	10,6±1,4
13	10	5	40	30	5	30	19,7±1,0
26	40	5	10	30	30	90	2,6±3,7
5	10	5	40	0	5	90	32,1±0,2
3	10	9	10	0	5	90	29,7±5,2
24	40	9	40	0	30	30	15,8±0,1
6	40	5	40	0	5	30	19,1±0,4
8	40	9	40	0	5	90	38,9±4,9
9	10	5	10	30	5	90	0,0±0,0
23	10	9	40	0	30	90	34,8±5,2

2	40	5	10	0	5	90	4,2±5,9
11	10	9	10	30	5	30	5,6±0,0
14	40	5	40	30	5	90	22,2±3,2
25	10	5	10	30	30	30	13,9±1,0
4	40	9	10	0	5	30	1,7±0,8
21	10	5	40	0	30	30	15,3±1,2
18	40	5	10	0	30	30	12,4±2,9
1	10	5	10	0	5	30	18,9±1,3
15	10	9	40	30	5	90	0,0±0,0
33	25	7	25	15	17,5	60	7,5±2,2
34	25	7	25	15	17,5	60	11,5±0,6
35	25	7	25	15	17,5	60	8,4±0,8

[1]Fast stirring time [2]Slow stirring time

To determine the influence of the factors on color removal, a significance level of 95% was set (p=0.05). In the Pareto diagram (Figure 25), which shows the effect of the factors studied, it can be seen that an increase in extract dosage (3) or pH (2) will have a positive effect on color removal. On the other hand, an increase in the initial salt concentration (4) and an increase in the initial AM concentration (1) will make it more difficult to treat the effluent. Among these variables, the dosage of the extract is the most influential variable, with any variation becoming significant according to the 95% statistical significance threshold. Although the influence of pH was not significant at the chosen statistical limit, previous studies show the importance of evaluating this variable for each pollutant separately [63].

The design presented can be described by a model that includes third-order interactions (Equation 3), whose R^2 value was 0.96405 and the lack of fit was negligible. The names X1 to X6 correspond to the names of the variables in Table 22.

% removal AM =16.9 - 7.0X + 6.5X2+ 12.5X3 - 5.4X4 + 1.OX5 - 1.4X6 -
2,1X1X2+ 3,4X1X3+ 8,2X1X4 - 1,6X1X5+ 2,1X1X6+1,2X2X3 - 6,1X2X4 +
4,1X2X5+ 2,8X2X6+ 3,2X3X4- 1,1X3X5+ 2,8X3X6+ 9,6X4X5- 2,5X4X6 -
1,8X5X6+ 5,9X1X2X3+ 4,4X1X2X4- 0,4X1X2X5+ 5,9X1X2X6- 0,1X1X3X4 -
3.4X1X3X5+ 3.4X1X3X6- 3.6X1X4X5+ 2.4X1X4X6- 1.4X1X5X6 (3)

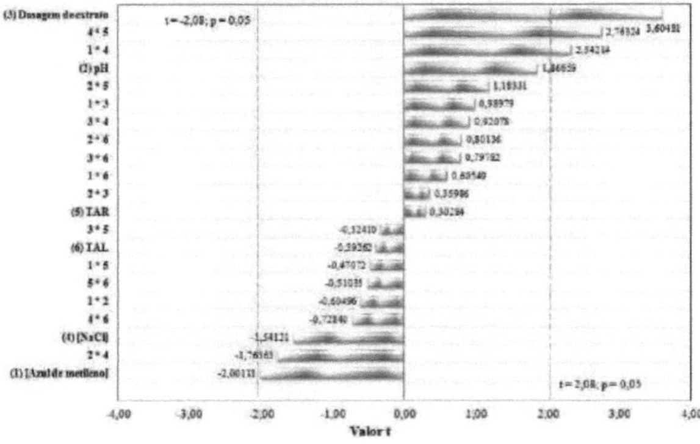

Figure 25. Pareto diagram of the first and second order effects on AM removal.

[1]A p-value = ±0.05 equals t=±2.084

As for the initial dye concentration, previous research shows that an increase in this factor negatively affects removal, more acutely for protein-based coagulants than for tannin-based coagulants [107]. Other studies have obtained greater removal efficiency at high initial dye concentrations, as is the case with *Moringa oleifera* seeds [63].

As for the times of the stirring stages, their variation does not have a significant influence on the percentage of final color and, therefore, it is concluded that the shortest times used, 5 and 30 minutes for the fast and slow stirring stages respectively, are sufficient to guarantee the process of contact and agglomeration of particles.

As far as the second-order effects are concerned, there is a significant positive effect of the interaction between AM concentration and salt concentration (factors 1*4), whose positive value indicates that an increase in these variables favors removal. A similar behavior is found in the second-order effect salt concentration - rapid stirring time (factors 4*5), since the presence of salt in the synthetic effluent can be eliminated by using longer rapid stirring times (30 minutes).

Figure 26 illustrates the behavior of the AM, salt and TAR concentrations on the response surface. It can be seen that the main effects of these variables show maximum percentage removal at low concentrations of AM, salt and TAR.

As for pH and TAR, the interaction of these factors with the addition of salt indicates a clear disadvantage in the removal of the dye, becoming detrimental in optimizing pH and affecting the sedimentation speed of the flocs formed. In general, the addition of salt hinders the decolorization process, a common phenomenon in the treatment of textile effluents in the dyeing stage [22]. The other second-order effects were not significant, with p-values of less than 0.05.

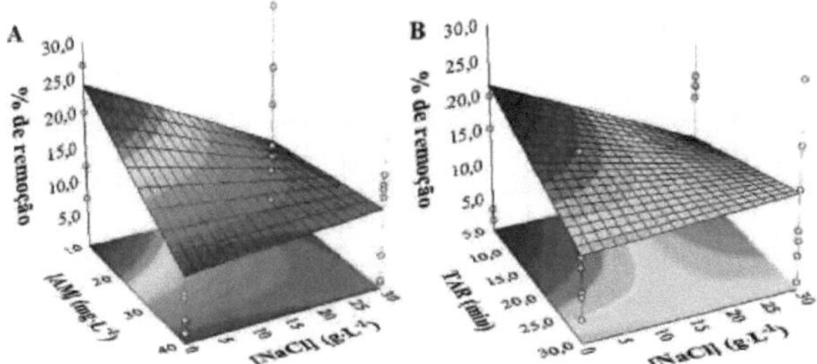

Figure 26. Response surface for AM decolorization for second-order interactions. A) Initial dye concentration and NaCl concentration. B) TAR and NaCl concentration.

Bearing in mind that pH and coagulant dosage were the factors that most influenced discoloration, and that the maximum removal percentages were achieved at the upper limits of the ranges for each of these variables, pH values of 10 and 11 and extract dosages higher than those originally established were tested. The other factors (TAR, TAL, [AM]INITIAL, [NaCl]) were considered parameters for all subsequent tests.

Figure 27 shows the effect of pH on AM decolorization. An increase in decolorization is observed at basic pHs, suggesting that the process of neutralizing the positive charges of the dye is favored, leading to the transformation of soluble pollutants into a precipitated solid phase [102]. The low removals at acidic pHs can be explained by the inhibition of the coagulating activity of the active ingredients studied: the presence of tannins suggests the establishment of hydrophobic bonds and hydrogen bridges, which generate complexes which, below pH 3.5, become unstable and dissociate [5]. On the other hand, the binding activity of protein compounds is favored at neutral or basic values, with a slight decrease in the latter. Acidic pH values (below 4) can completely inhibit this capacity. [50].

Due to the mixed nature of the extracts obtained, it is not possible to determine the participation of each active ingredient in the coagulation-flocculation process. Only with the pure compounds is it possible to determine the role of each component.

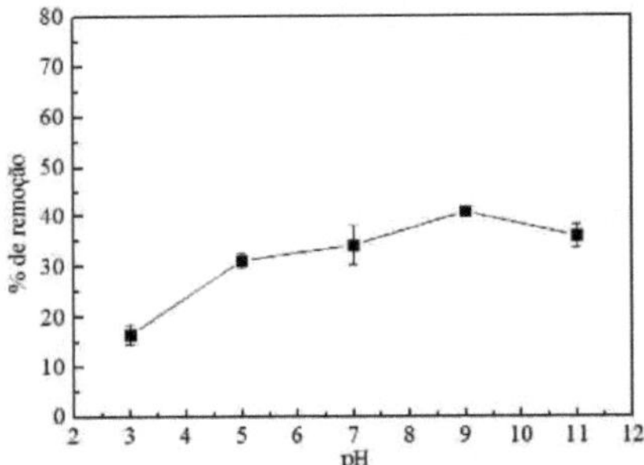

Figura 27. AM discoloration as a function of pH. Parameters: [AM]INicIAL=10 mg· L"1, [NaCl]=0 g· L(-1), TAR=5 min, TAL = 30 min, extract dosage=40 ml· L(-1).

Once the pH value had been established, higher dosages of extract than those used in the experimental design were tested. To determine the appropriate dosage, it is necessary to know the coagulating capacity (***qc***), which is defined as the amount of dye removed by a given volume of coagulant (mg of dye· ml of extract-1) according to Equation 4 [16]:

$$q_C = \frac{(C_0 - C_i) * V_C}{V_E} \quad (4)$$

Where ***Co*** is the initial dye concentration (mg· L(-)1), ***Ci*** *is* the concentration reached after the established sedimentation period (mg· L(-1)), ***VC*** is the final volume of dye solution used (L) and ***VE*** is the extract dosage (ml· L(-1)).

Figure 28 shows the variation in the percentage of color removal as a function of the extract dosage. An increase in color removal is evident as the extract dosage increases, a common behavior in the use of natural coagulants [100]. However, a constant decrease in coagulant capacity can be observed, and the presence of the extract's own colorations can be seen from dosages of 60 ml· L(-1). It was decided to work with an intermediate dosage that offers acceptable removal while still having a good coagulating capacity, in the case of AM, 40 ml· L(-1).

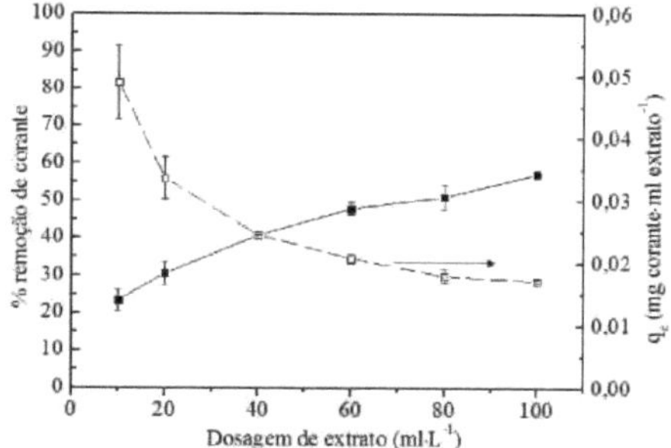

Figura 28. AM discoloration as a function of extract dosage. Parameters: [AM]INITIAL=10 mg· L(-1), [NaCl]=0 g· L(-1), TAR=5 min, TAL 30 min, pH=9.

In order to establish a comparison of the dosages used in this work with other studies [64,100,128], equation 5 shows the value of tannins dosed by weight, calculated from the concentration of total tannins in extract E5 (Table 15). A dosage of 40 ml· L(-1) is equivalent to 12.4 mg of tannin per liter of dye solution; common dosages of natural coagulants (*Moringa Oleifera* and Acacia Negra) are in the range of 0.5 to 500 mg· L(-1) [128].

$$\text{Dosagem de extrato} = \frac{40\ ml_{Extrato\ concentrado}}{1\ L} * \frac{1\ ml_{Extrato\ bruto}}{0{,}15\ ml_{Extrato\ concentrado}} * \frac{1\ g_{Caule}}{20\ ml_{Extrato\ bruto}} * \frac{0{,}93\ mg_{Taninos}}{1\ g_{Caule}} = 12{,}4 mg_{Tanino} \cdot L_{Efluente}{}^{-1} \quad (5)$$

The percentages of dye removal were determined from measurements taken after the samples had rested (60 minutes). It is known that the coagulation process occurs rapidly, and some authors [63] suggest that the equilibrium concentration of the dye is reached in the rapid agitation stage, so it is likely that complex coagulation mechanisms involving the formation of a network-like structure do not require a long contact time. This is an advantage compared to other processes such as adsorption, where contact times are longer.

With regard to the type of dye, both the chromophore group and the auxochrome group present in the dye molecule play an important role during the coagulation-flocculation process. The small chains and low charge, in this case cationic for AM, can negatively influence the process [128].

Regarding the action of the crude extract in the decolorization of AM solutions, it is believed that the phenomena of complex formation and charge neutralization by affinity between pollutants and coagulants are the main mechanisms proposed [149]. In the case of the removal of cationic compounds, the theory is proposed that the various OH groups present in the natural polyphenolic extract can interact with the cationic dye via ion-ion and ion-dipole forces. The anionic character of natural polyelectrolytes, such as those extracted from the Nirmali tree (*Strychnos potatorum),* originally from India, has been explained in the literature by the presence of hydroxyl (OH^-) and carboxyl (COO^-) groups [119]. In addition, the nature of condensed tannins suggests the possibility of forming non-covalent interactions with organic compounds, such as AM in this case [12].

The following sections present the results of the decolorization tests for two other types of cationic dyes, in order to evaluate the behavior of the crude extract for different molecular structures. For these dyes, only the results relating to variations in pH and extract dosage will be presented, factors considered to be the most positively significant according to the experimental design carried out for AM.

1.5.2. Blue Maxilon GRL 300% (AMG)

Figure 29 and Figure 30 show the influence of pH and extract dosage on solutions of AMG, an azo dye. Previous studies indicate that this type of dye is easier to remove color from, suggesting that the azo group is the chromophore that is easier to coagulate [100,128]. In this work, the highest efficiencies (without the use of an alkalizing agent) were achieved for AMG. The number of azo groups influences removal, with a smaller number of such bonds leading to greater efficiency, as is the case with AMG, which is a monoazo dye. The increase in the decolorization of solutions containing this type of dye is also related to the high degree of dissociation of aqueous solutions of AMG and the dye's weak aggregation mechanisms [107].

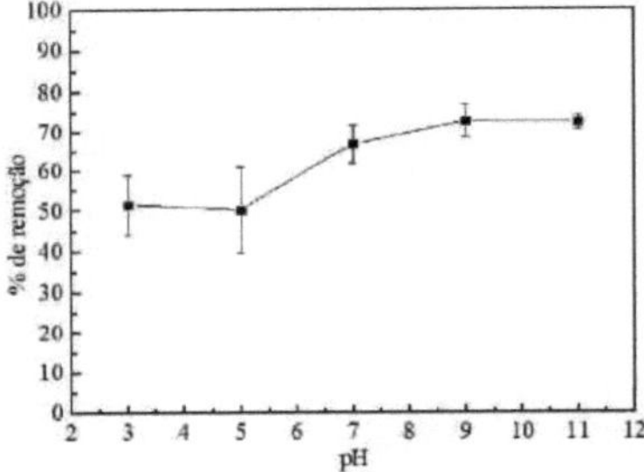

Figura 29. AMG discoloration as a function of pH. Parameters: $[AMG]_{INITIAL}$=10 mg·$^{L(-1)}$, [NaCl]=0 g·$^{L(-1)}$, TAR=5 min, TAL = 30 min, extract dosage=40 ml g·$^{L(-1)}$.

One of the problems encountered in the textile industry is the fluctuation of effluent parameters, especially pH, color and chemical oxygen demand, due to the nature of the process (batch operation) [128]. As the pH increases, the number of positively charged sites in the medium decreases and a greater negative charge favors electrostatic attractions between the cationic dye and the coagulant [132]. However, in the case of AMG, basic pH values do not show significant differences with removals carried out at neutral pH, which is why pH 7 was determined as the optimum value. This system therefore shows greater flexibility with regard to the pH value of the extract than in other studies, where a change in pH is highly significant [63,150]. Similarly, the action of the extract at a neutral pH suggests that the formation of bridges between molecules, a mechanism in which the charge of the particles is of secondary importance, prevails over the neutralization of charges, a phenomenon described in the literature [127].

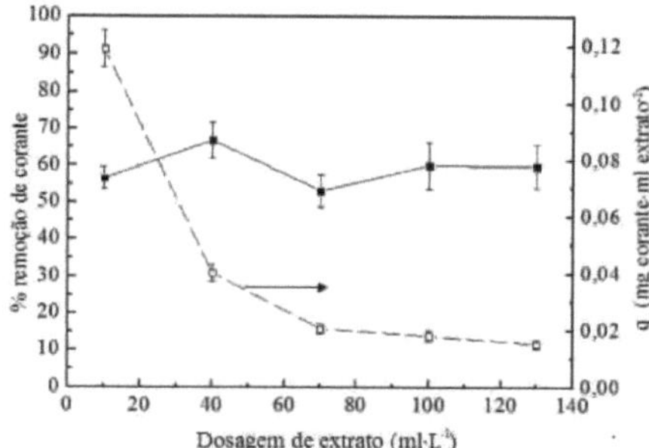

Figura 30. AMG discoloration as a function of extract dosage. Parameters: $[AMG]_{INIC1}A_L$=10 mg· $^{L(-1)}$, [NaCl]=0 g· $^{L(-1)}$, TAR=5 min, TAL=30 min, pH=7.

Since part of the proposed mechanism for coagulation is the adsorption of colloids and dissolved dye molecules on the surface of the flocs already formed, the zero charge point (pH_{CZ}) can help describe the decolorization phenomenon by describing how changes in pH affect the process. The interactions of cations with the surface of the material are favored at pH>pHCZ [104]; the value of the zero charge point found was 6.2 (APPENDIX 1), thus justifying the increase in removal observed in the pH 5 to 7 range, which can be explained by the affinity of the surface to the dye at pH values above pHCZ.

Evaluating the extract dosage, slight increases in color removal are observed with the increase, however there is a rapid decrease in coagulating capacity. The increase in dosage is significant up to values of around 40 ml· $^{L(-1)}$, the maximum removal value found. Previous research suggests that the decrease in removal after a maximum point is due to a restabilization of the suspension by overdosing, reducing the efficiency of the process [99]. In Figure 30 it can also be seen that the changes in removal with increasing extract dosage are much smaller than in the case of AM (Figure 28).

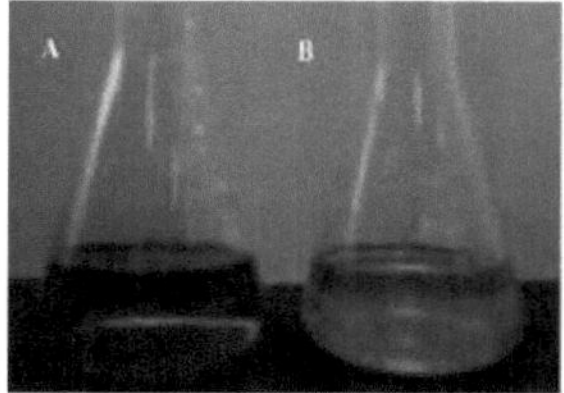

Figura 31. AMG samples. A) Original solution and B) solution with added extract. Conditions: $[AMG]_{INIC1}A_L$=10 mg· $^{L(-1)}$, [NaCl]=0 g· $^{L(-1)}$, TAR=5 min, TAL=30 min, pH=7, extract dosage=40 ml· $^{L(-1)}$.

1.5.3. Malachite green (MV)

Figure 32 shows the removals obtained for different dosages and pH values. For this dye, decolorization percentages of around 60 to 80% were achieved, values similar to those presented by Bibi et al. [131], who used phenolic extracts obtained from peanut seeds and shells, achieving removal percentages of between 70 and 90% by varying the dosage of the extract and monitoring the incubation time.

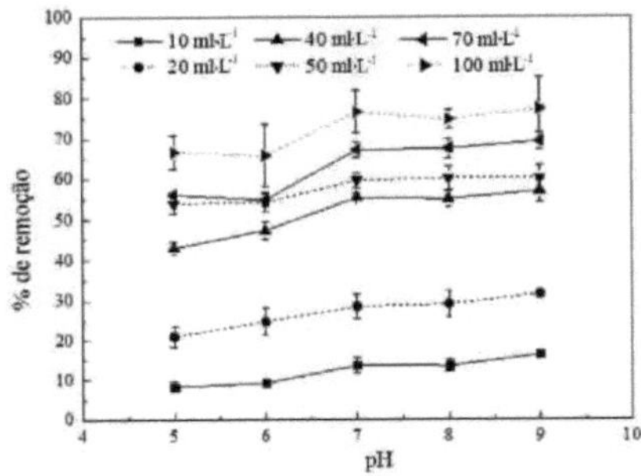

Figura 32. VM discoloration as a function of pH at different extract dosages. Parameters: $[VM]_{INITIAL}$=10 mg·$^{L(-1)}$, [NaCl]=0 g·$^{L(-1)}$, TAR=5 min, TAL = 30 min.

In some cases, the pH of the dye solution can affect the original structure of the dye, as is the case with VM, reducing the intensity of the color in the effluent. It has been reported that the color of a VM solution is stable in the pH range of 3-7, and values above this range can lead to erroneous conclusions [151]. Therefore, considering the above results, a neutral pH was chosen as the working parameter, ensuring that the discoloration was caused by the addition of the extract and not by molecular changes in the dye due to the change in pH. Figure 32 shows a practically stable behavior at basic pH, with no significant increase in discoloration. This behavior has been observed in the literature [104]. This behavior suggests, as in the case of AMG, the presence of hydrogen bridge formation mechanisms, especially at pHs close to 7.

The results obtained for VM differ from other studies [12], which report significant decreases in removal at basic pH, and which could be related to a greater availability of negatively charged sites in the extract, promoting charge neutralization phenomena and also a possible degradation of the dye during the incubation period. The difficulty in removing this dye at acidic pH has been reported in the literature. These removals, when compared to those of azo or anthraquinonic dyes, triarylmethanes present greater difficulty and lower removal percentages are observed [128].

The effect of the extract dosage in the MV solutions can be seen in Figure 33, in which dosages above 50 ml·$^{L(-1)}$ showed difficulties in sedimenting the flocs, possibly due to the extract being overdosed, forming layers around the suspended solids and having a negative effect on decolorization [110]. In addition, dosages higher than 60 ml·$^{L-1}$ interfere with the evaluation of color removal, as the extract colors the solution.

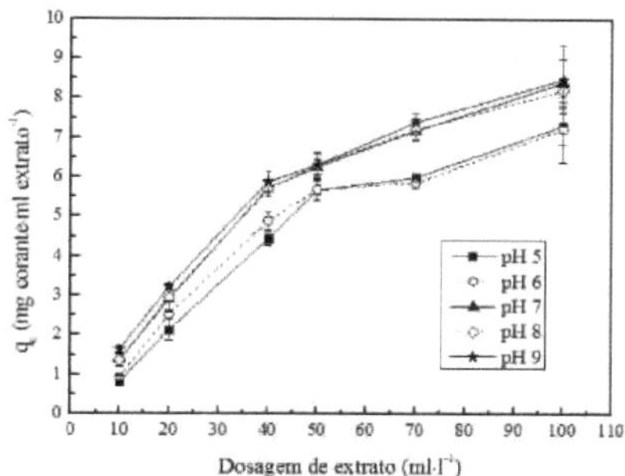

Figura 33. Coagulating capacity of the extract in VM solutions as a function of extract dosage. Parameters: [VM]INITIAL=10 mg· $^{L(-1)}$, [NaCl]=0 g· $^{L(-1)}$, TAR=5 min, TAL = 30 min.

In the VM samples, floc formation was slower than in the other dyes tested, requiring longer times for total sedimentation of the agglomerated particles. The structure of the dye, with a low cationic charge, may have influenced the decrease in coagulant activity [128]. It was observed that the flocs formed were light and easily dispersible, which suggests that just adding the extract for this treatment may not be enough to completely destabilize the dye molecules. Figure 34 shows the result of the decolorization test under the optimum conditions found.

Figura 34. Samples of VM. A) Original solution and B) With addition of extract. Conditions: $[VM]_{INITIAL}$=10 mg· $^{L(-1)}$, [NaCl]=0 g· $^{L(-1)}$, TAR=5 min, TAL=30 min, pH=7, extract dosage=50 ml· $^{L(-1)}$.

After evaluating the action of the E5 extract on the different dyes, we can see the importance of studying the working conditions separately in order to achieve better decolorization, especially with regard to determining the appropriate pH and extract dosage.

1.5.4. Addition of alkalizing agent

With the pH conditions and extract dosage established, the addition of sodium carbonate as an alkalizer was tested, evaluating its effect on color removal and the stability of the flocs formed. The simultaneous application of the two agents, extract and alkalizer, in the rapid stirring stage, ensured homogenization of the dye solution, an essential requirement for the formation of flocs. The effects of adding carbonate are shown in Figure 35. It can be seen that there is a carbonate dosage at which removal is maximum for AM and AMG. The dosages used in this work are similar to those used in research using carbonate associated with aluminum salts as coagulants [22].

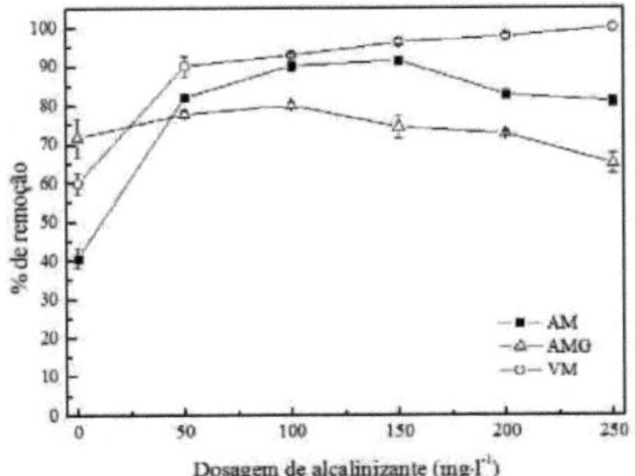

Figura 35. Discoloration tests using sodium carbonate as an alkalizing agent. Parameters: $[Dye]_{INITIAL}$=10 mg· L(-1), [NaCl]=0 g· L(-1), TAR=5 min, TAL = 30 min. pH values and extract dosage according to ά Table 23 for each dye.

For VM, carbonate dosages above 50 mg· mL(-1) lead to slightly higher removals, but the observed stability of the flocs formed was very low, and they remained in suspension even after 48 h of sample incubation. With a view to a larger-scale application that could only remove the sedimented dye with incubation time, a dosage of 50 mg· mL(-1) of carbonate was chosen.

The presence of an alkalizing agent helps in the generation and subsequent settling of flocs [22]. In this situation, a greater availability of negatively charged ions increases the surface available for agglomerating particles and favors an increase in the size of the particles suspended in the adsorption process. Table 23 shows the increase in removal efficiency with the use of the alkalizing agent for the three colorants tested. In the cases of AM (Figure 36) and VM, the action of the alkalizer considerably increased the removal percentages. For AMG, the effect was negligible. In the case of the dye samples treated with alkalizer alone, there was no color difference with the initial dye solutions, nor was there any floc formation.

Table 23. Optimum removal conditions using the aqueous phase concentrated ethanolic extract of cassava stems (E5).

Parameters	**Dye**[1]		
	AM	**AMG**	**VM**
Extract dosage (ml·L-1)	40	40	50
NaCl (g·L-1)	0	0	0
Alkalizing ($mg \cdot L^{-1}$)	150	100	50
TAR (min)[2]	5	5	5
TAL (min)[3]	30	30	30
Color removal without alkalizing (%)[4]	40,7	71,8	64,8
Color removal with alkalizing agent (%)[4]	91,4	80,0	90,1
Original pH of the dye solution	6,1	6,0	5,0
pH adjusted	9,0	7,0	7,0
Final pH of treated dye solution without alkalizing agent	6,0	5,8	5,5
Final pH of dye solution treated with alkalizing agent	6,9	7,0	7,0

[1]Dye concentration: 10 mg·$^{L(-1)}$.

[2]Speed in the rapid stirring stage: 170 rpm.

[3]Slow stirring speed: 50 rpm.

[4]Calculated removal of samples with a sedimentation time of 1 hour

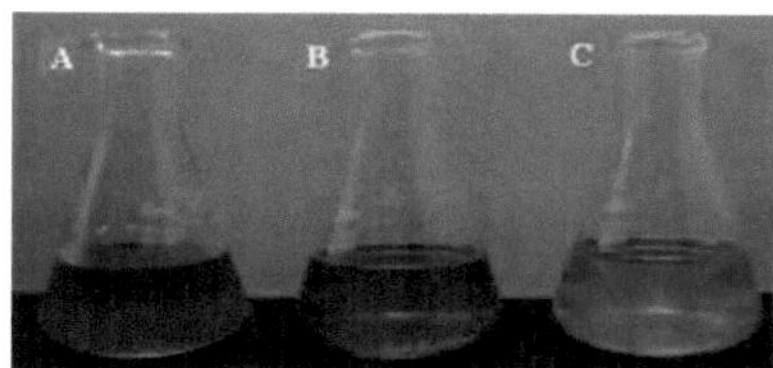

Figura 36. Samples of AM (10 mg·$^{L(-1)}$). A) Without treatment. B) With crude extract. C) With crude extract and alkalizing agent. Photographs taken after 24 hours of incubation at 25 °C with prior removal of the flocs formed. Working conditions according to Table 23

In the solutions treated without the addition of an alkalizing agent, there was a decrease in the final pH compared to that adjusted for the tests. The variation may be related to the low ionic strength of the initial solution (distilled water without any buffering effect), and also to the slight acidity of the polyphenolic extract (pH between 5 and 6). Complementary tests containing buffered solutions may show greater pH stability during the process. In the samples submitted to the addition of an alkalizing agent, the final pH (after treatment) was close to 7, with no need to readjust it, an advantage already documented for the use of carbonate together with aluminium salts [17,124]. The alkalizing agent can also act as a neutralizer of acidic substances formed in the medium [104] or already present in the crude extract.

The conditions in Table 23 were used to evaluate the chemical oxygen demand (COD) and the toxicity of the untreated and treated solutions.

5.6 Determination of COD

Table 24 summarizes the results of the COD determination for the treated dye solutions after total removal of the flocs formed. The initial COD values are common for industrial effluents containing dyes (paper production, textiles, etc.), which are in the order of 10-80 mgO2·$^{L(-1)}$ [128].

There is a clear increase in the amount of organic matter present in the solutions tested with the natural extract, and this increase is proportional to the increase in the dosage of the extract. This is a common phenomenon in effluents treated with natural coagulants. Anastasakis et al. [114] present various types of natural flocculants, which increase the organic load (measured as dissolved organic carbon - DOC) of the treated effluent by different percentages. Okuda et al. [62] determined that the increase in organic load using aqueous extracts of *Moringa oleifera* is mainly due to inactive components present in the extract.

There is no clear trend in the variation of COD with the addition of alkalizing agent, suggesting a non-significant effect on this variable when carbonate is added.

Table 24. COD of treated dye solutions

Dye	COD dye solution	Extract dosage (ml- L^{-1})			
		40	60	40	60
		COD Without alkalizing		COD With alkalizing [1]	
AM	105	1134	1938	1084	1502
AMG	53	991	1070	962	1143
VM	59	1038	1555	1423	1885

[1]Dosage of alkalizing agent according to table 23. [2]COD values in $mgO_2 \cdot L^{(-)}1$

In this study, the extracts were not previously treated. For further research, delipidation processes [62], the removal of pigments or the addition of antioxidants [89] could be applied with the aim of eliminating organic load or compounds present in the extract that could interfere with the final COD of the treated effluent. The removal of fatty acids, sugars or pigments does not affect the removal efficiency, which is specifically related to the presence of polyphenols in the extract [12]. Specific analytical techniques (gas chromatography-mass spectrometry) could help determine the various organic compounds present in the extract.

The use of natural compounds with a secondary coagulating or flocculating action is also an interesting alternative for not affecting the organic load [105]. It is important to note that the increase in organic load must be monitored in all cases, due to the possibility of toxic chlorinated products being formed in potential subsequent chlorination steps for the effluent in question [114]. The use of biological treatments with the aim of degrading organic matter is necessary according to the final conditions of the solutions tested.

5.7 Toxicity tests

The toxicity test with *Artemia salina* was carried out under three different conditions:

- Dye solution 10 $mg \cdot L^{(-1)}$
- Dye solution treated with extract under the conditions set out in Table 23 without removal of the flocs formed.
- Dye solution treated with extract under the conditions set out in Table 23 with removal of the flocs formed (clarified effluent).

Table 25 shows the results obtained, expressed as the mortality (in percentage) of the *Artemia salina* population for each condition evaluated. For the dye solutions without any type of treatment, the mortality rate was low compared to previous studies which describe lethal dosages for cationic dyes of around 1 $mg \cdot L^{(-1)}$ [130,131].

Table 25. Toxicity of dye solutions before and after treatment with ethanolic extract of cassava stems.

Dye[1]	Mortality (%)		
	Initial solution	After adding extract[2]	After adding extract and removing agglomerate[1]
AM	18,5±4,5	72,5±9,6	12,4±4,3
AMG	18,7±1,6	80,0±8,2	12,4±5,0
VM	24,7±5,8	63,8±11,1	3,6±4,3

[1]Dye concentration: 10 $mg \cdot L^{(-1.)}$

[2]Process parameters according to table 23.

In the treated effluents, there was a significant difference in the samples in which the agglomerated material was not removed but remained sedimented or suspended. In these samples, mortality increased compared to the untreated solutions; the increase in COD depressing the amount of O_2 and the possible formation of toxic compounds over the incubation period (24 h) may explain this increase. In the samples in which the coagulated material was removed and the solution was clarified, the toxicity decreased, showing lower percentages than the initial solutions. Therefore, the elimination of the molecular complexes formed during treatment is necessary to obtain an effluent that is less harmful to the aquatic environment.

To assess the toxicity of the extract, the lethal concentration (LC50) was determined, which causes the death of 50% of the organisms present in the aquatic environment. Figure 37 shows the curve obtained for the mortality of microcrustaceans as a function of the concentration of extract in water, in which case concentrations above 80 ml of extract· $^{L(-)}$1 are considered lethal for the aquatic environment studied.

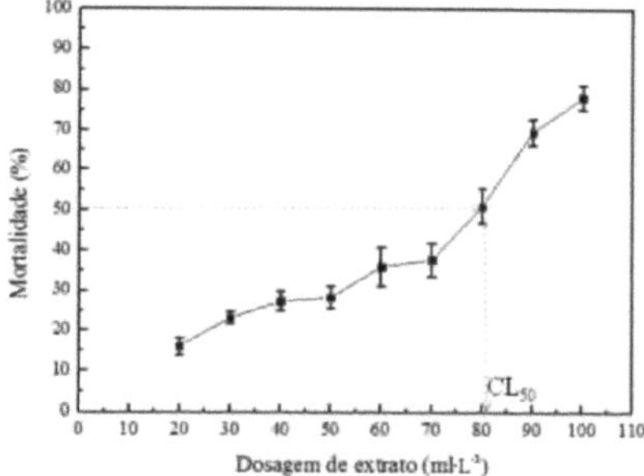

Figure 37. Toxicity test with *Artemia salina* of the aqueous fraction of the concentrated ethanolic extract of the stem (E5). Determination of CL50

The literature reports that natural anionic or non-ionic coagulants are generally low in toxicity, while cationic polyelectrolytes are more toxic, especially for aquatic organisms [102]. On the other hand, most commercial petroleum-based polymers used in water treatment are not entirely safe or environmentally friendly. This has led to growing interest in developing natural, low-cost alternatives [99].

6 CONCLUSIONS

This study determined the optimum extraction conditions for favoring the hemagglutinating activity of aqueous extracts of cassava leaves: temperature of 25°C and times between 45 and 60 min. Storage at a low temperature (10°C) preserved the HA of the extracts for up to 3 days.

HA is favored by purification processes such as isoelectric precipitation, allowing agglutination to be promoted with a smaller amount of protein. The increase in temperature and extraction time favors an increase in the percentage of proteins obtained.

A selective character of the lectin present in the cassava leaf was observed, taking into account the different AH values for each type of erythrocyte. No activity was detected in pig blood samples.

The protein and phenolic compound contents in the crude extracts were higher in the leaves than in the stem in all the extractions made. The percentage of condensed tannins extracted with ethanol was higher than that observed for aqueous extraction.

Comparing the levels of polyphenols and tannins obtained in extractions made with 70% acetone, the recovery with 70% ethanol was 60% while for the aqueous extract it was 27%.

The aqueous phase of the concentrated ethanolic extract of the cassava stem was the most suitable fraction for the decolorization tests, achieving significant percentages of removal. The major presence of condensed tannins suggests their importance in the coagulation phenomenon in subsequent decolorization tests.

The FTIR spectra made it possible to identify the main functional groups present in the extracts, helping to propose molecular structures that are part of the agglomeration processes in the HA and decolorization tests. The spectra obtained show the characteristic peaks of protein compounds (dialyzed aqueous extract) and polyphenolic compounds (freeze-dried ethanolic extract).

The mechanisms proposed as promoting coagulation-flocculation, such as molecular interactions, floc formation and subsequent precipitation, are caused by complex physico-chemical processes that develop progressively during the sample's incubation period.

In the decolorization tests, extract dosage and pH were determined to be the main and most influential factors in color removal efficiency. Stirring times in the slow and fast stages did not affect the extract's performance. The initial concentration of dye and the dosage of NaCl affected the process negatively, which is why the minimum quantities evaluated in the factorial design were set.

The optimum extract dosages were around 40 ml$\cdot^{L(-)}$1 and the pH values with the best removals were 9 for AM and 7 for AMG and VM.

The extracts used showed acceptable removal percentages when compared to natural coagulants from other plant sources.

The slow formation of flakes can be explained by inefficient dye-extract binding. Low molecular weights in the active ingredients (tannins and proteins) and the molecular structural non-linearity of the polyphenols are related to this process.

The contents of active agents (proteins and polyphenols) determined in the stem are lower than those found in plant sources commonly used for effluent treatment (*Moringa Oleifera)* and, consequently, the volumes of extract dosed are high.

The addition of an alkalizing agent increased the efficiency of the process. Optimum dosages of 150 and 100 mg· $^{L(-1)}$ were found for AM and AMG respectively. For VM, the addition of carbonate at dosages above 50 mg· $^{L(-1)}$ led to the re-stabilization of the suspended material, not allowing the natural sedimentation of the flocs formed.

There was a clear increase in the COD of the solutions treated with the extracts, due to the presence of additional active organic components to the active ingredients. The addition of the alkalizing agent did not generate any changes in the COD of the dye solutions, improving the efficiency of the process.

There was a decrease in toxicity in the samples tested and the extract dosages used were lower than the LD50 (80 ml· $^{L(-1)}$).

The simplicity of coagulation-flocculation techniques makes the process with natural coagulants cheaper and more attractive. The potential use of the aerial part of cassava in effluent treatment takes into account the presence of active principles that promote the coagulation phenomenon. Its use is linked to the use of suitable extraction processes with significant efficiencies in the extraction of proteins and phenolic compounds.

7 SUGGESTIONS

Based on the results obtained, the following topics are suggested for future work:

- Carry out bleaching tests with extracts obtained from older plants (12 months) in order to obtain a higher percentage of condensed tannins.
- Evaluate the yield of protein, phenolics and tannins in the final crude extracts by making successive extractions of the solid tested.
- Use ethanolic extracts with a higher concentration than the one used in this study (15% of the initial volume of the extract) for the bleaching tests.
- Carry out a study of the color removal efficiency of saline extracts of the leaves and stem and compare the results with aqueous and ethanolic extractions.
- Evaluate the use of advanced characterization techniques to determine the functional groups present in the extracts.
- To characterize the crude extracts in terms of lipid and carbohydrate composition and their influence on the increase in COD in the treated effluents.
- Monitor discoloration as a function of the incubation period in order to study the sedimentation rate of the flocs formed.
- Carry out appropriate studies on a pilot scale to assess the extract's performance capacity and establish realistic costs and the potential for using the product on an industrial scale.

REFERENCES

1. PANDEY, A. et al. Biotechnological potential of agro-industrial residues. II: cassava bagasse. **Bioresource Technology**, v. 74, p. 81-87, 2000.

2. FASUYI, A. O.; ALETOR, V. A. Varietal composition and functional properties of Cassava (*Manihot Esculenta* Crantz) leaf meal and leaf protein concentrates. **Pakistan Journal of Nutrition**, v. 4, n. 1, p. 43-49, 2005.

3. WOBETO, C. et al. Antinutrients in the Cassava (*Manihot esculenta* Crantz) leaf powder at three ages of the plant. **Ciência e Tecnologia de Alimentos**, v. 27, n. 1, p. 108-112, 2007.

4. GOLDSTEIN, I. J. et al. What should be called a lectin? **Nature**, v. 66, p. 285, 1980.

5. MIN, B. R. et al. The effect of condensed tannins on the nutrition and health of rumiants fed fresh temperate forages: a review. **Animal Feed Science and Technology**, v. 106, p. 3-19, 2003.

6. SANTOS, A. D. F. S. **Bioactive molecules from *Moringa Oleifera*: detection, characterization and isolation**. 2007. 112 f. Thesis (Doctorate in Biological Sciences) - Federal University of Pernambuco. Recife, PE. 2007.

7. OZACAR, M.; SENGIL, I. A. The use of tannins from turkish acorns (Valonia) in water treatment as a coagulant and coagulant aid. **Turkish Journal of Engineering and Environmental Science**, v. 26, p. 255-263, 2002.

8. MELO, D. S. D. **Cassava leaf meal: effects on peroxidation and lipid profile in the liver and blood of rats.** 2005. 74 f. Dissertation (Master's Degree in Agrochemistry) - Federal University of Lavras. Lavras, MG. 2005.

9. CORREA, D. A. et al. Removal of polyphenols from cassava leaf flour. **Ciência e Tecnologia de Alimentos**, Campinas, v. 24, n. 2, p. 159-164, 2004.

10. SILVA, M. C. **Partial characterization of a lectin from cassava leaves (*Manihot esculenta* Crantz)**. 2008. 55 f. Dissertation (master's degree in Agrochemistry) - Federal University of Lavras. Lavras, MG. 2008.

11. MARKOM, M. et al. Extraction of hydrolysable tannins from Phyllanthus niruri Linn.: Effects of solvents and extraction methods. **Separation and Purification Technology**, v. 52, p. 487496, 2007.

12. JEON, J. et al. Use of grape seed and its natural polyphenol extracts as a natural organic coagulant for removal of cationic dyes. **Chemosphere**, v. 77, p. 1090-1098, 2009.

13. PEREIRA, C. A. et al. Hemagglutinin of cassava leaves (*Manihot esculenta* Crantz): partial purification and toxicity. **Ciência e Agrotecnologia de Lavras**, v. 32, n. 3, p. 900-907, 2008.

14. SANTOS, A. F. S. et al. Isolation of a seed coagulant *Moringa oleifera* lectin. **Process Biochemistry**, v. 44, p. 504-508, 2009.

15. BELTRAN-HEREDIA, J.; SANCHEZ MARTiN, J.; SOLERA- HERNANDEZ, C. Removal of sodium

dodecyl benzene sulfonate from water by means of a new tannin-based coagulant: Optimization studies through design of experiments. **Chemical Engineering Journal**, v. 153, p. 56-61, 2009.

16. BELTRAN HEREDIA, J.; SANCHEZ-MARTiN, J.; MARTiN- SANCHEZ, C. Remediation of dye-polluted solutions by a new tannin-based coagulant. **Industrial & Engineering Chemistry Research**, v. 50, p. 686-693, 2011.

17. MRWA: MINNESOTA RURAL WATER ASSOCIATION. Coagulation and Flocculation Process Fundamentals, 2005. Available at: <www.mrwa.com/OP-Coagulation.pdf >. Accessed on: August 22, 2011.

18. IITK - INDIAN INSTITUTE OF TECHNOLOGY KANPUR. Water & Wastewater Engineering, 2010. Available at: <http://nptel.iitm.ac.in/courses/Webcourse-contents/IIT-KANPUR/wasteWater/Double%20Layer%20Compression.htm>. Accessed on: Aug. 22, 2011.

19. OZACAR, M.; SENGIL, A. Efectiveness of tannins obtained from Valonia as a coagulant aid for dewatering of sludge. **Water Research**, v. 34, n. 4, p. 1407-1412, 2000.

20. MILLER, S. M. et al. Toward understading the efficacy and mechanism of *Opuntia spp.* as a natural coagulant for potential application in water treatment. **Environmental Science and Technology**, v. 42, p. 4274-4279, 2008.

21. ROYER, B. **Removal of textile dyes using *Araucaria angustifolia* seed bark as a biosorbent**. 2008. 68 f. Dissertation (Master's Degree in Chemistry) - Federal University of Rio Grande do Sul. Porto Alegre, RS. 2008.

22. FURLAN, F. R. **Evaluation of the efficiency of the coagulation-flocculation and adsorption process in the treatment of textile effluents**. 2008. 151 f. Dissertation (Master's Degree in Engineering

Chemistry) - Federal University of Santa Catarina. Florianopolis, SC 2008

23. STUPAK,M etal Biotechnologicalapproachestocassavaprotein improvement **Trends in Food Science & Technology**,v 17,p 634-641,2006

24. CABALLERO, B (Ed) **Encylopedia of Foods Science and Nutrition** Baltimore:AcademicPress ,2003

25. CEBALLOS,H etal Variationincrudeproteincontentincassava (*Manihot esculentaCrantz*)roots **Journal of Food Composition and Analysis**, v. 19, p . 589-593, 2006.

26. FAO FAO STATS, 2009 Available at: <http://faostat.fao.org/site/567/default.aspx#ancor>. Accessed on: July 7, 2011.

27. NETO,C.R.;MARCOLAN,A.L.**Exploratory study on the consumption behavior of cassava and its derivatives in Brazil, with emphasis on the Northern Region**. 48 SOBER Congress. Porto Velho: Sociedade Brasileira de Economia, Administração e SociologiaRural.25a28dejulhode2008.p.1-20.

28. GROXKO, M. **Cassava. Analysis of the agricultural situation. 2010/11 harvest**. Department of Agriculture and Supply. Department of Rural Economy. State of Paranà. Curitiba, PA. 2010.16p.

29. BOHNENBERGER, L. **Cassava leaf protein concentrate as a food supplement for Nile tilapia**. 2008. 68 f. Dissertation (Master's Degree in Agricultural Engineering) - State University of Western Paranà. Cascavel, PA, p. 68. 2008.

30. SILVA,J.L.D.**Obtença de concentrado proteico de folhas e parte aerea da mandioca (*Manihot esculenta* Crantz)**.2007.91 f. Dissertation (Master's Degree in Agricultural Engineering) - State University of Western Paranà. Cascavel, PA. 2007.

31. CEREDA,M.P.**Manejo, Uso e Tratamento de Subprodutos da Industrializaçao da Cassava**.SaoPaulo:FundaçaoCargill,v.4, 2001. 320 p. Latin American Tuberous Amylaceous Crops Series.

32. CARVALHO, V. D. D.; KATO, M. D. S. A. Potential use of the aerial part of cassava.**Informe Agropecuario**,v. 13,n.145,p.23-28,1987.

33. WOBETO,C.etal.NutrientsintheCassava(*Manihot esculenta* Crantz) leafmealatthreeagesoftheplant. **Ciência y Tecnologia de Alimentos**,v.26,n.4,p.865-869,2006.

34. MELO, D. S. D. et al. Effects of cassava leaf flour on lipid peroxidation, blood lipid profile and liver weight in rats. **Ciência e Agrotecnologia**, v. 31, n. 2, p. 420-428, 2007.

35. SGARBIERI, V. C. **Alimentaçâo e Nutriçâo:** fator de saúde e desenvolvimento. Sao Paulo: Almed, 1987. 387 p.

36. BUITRAGO, J.; GIL, J. C. La yuca en la alimentación animal. In: CEBALLOS, H.; OSPINA, B. **La yuca en el tercer milenio**. Cali: CIAT, 2002. Chap. 28, p. 531.

37. SHARON, N.; LIS, H. Legume lectins - a large family of homologous proteins. **FASEB Journal**, v. 4, n. 3, p. 198-208, 1990.

38. CHEVREUIL, L. R. et al. Activity of hemagglutinating proteins in seeds of arboreal legumes from the Amazon flora. **Revista Brasileira de Neurociências**, v. 5, n. 2, p. 1020-1022, 2007.

39. OCCENA, I. V.; MOJICA, E. E.; MERCA, F. E. Isolation and partial characterization of a lectin from the seeds of Artocarpus camansi Blanco. **Asian Journal of Plant Sciences**, v. 6, n. 5, p. 757-764, 2007.

40. FRANCO-FRAGUAS, L. et al. Preparative purification of soybean agglutinin by affinity chromatography and its immobilization for polysaccharide isolation. **Journal of Chromatography B**, v. 790, p. 365-372, 2003.

41. KENNEDY, J. F. et al. Lectins, vesatile proteins of recognition: a review. **Carbohydrate Polymers**, v. 26, p. 219-230, 1995.

42. VASCONCELOS, I. M.; OLIVEIRA, J. T. A. Antinutritional properties of plant lectins. **Toxicon**, Oxford, v. 44, n. 4, p. 385-403, 2004.

43. PUSZTAI, A. et al. Effects of an orally administered mistletoe (type-2 RIP) lectin on growth, body composition, small instetinal structure, and insulin levels in young rats. **Nutritional Biochemistry**, v. 9, p. 31-36, 1998.

44. FU, L. L. et al. Plant lectins: Targeting programmed cell death pathways as antitumor agents. **The International Journal of Biochemistry & Cell Biology**, v. 43, n. 10, p. 1442-1449, 2011

45. PILOBELLO, K. T. et al. Development of a lectin microarray for the rapid analysis of protein glycopartners. **ChemBioChem**, v. 6, p. 985-989, 2005.

46. FRANCO-FRAGUAS, L. F.; BATISTA-VIEIRA, F.; CARLSSON, J. Preparation of high-density Concanavalin A adsorbent and its use for rapid, high-yield purification of peroxidase from horseradish roots. **Journal of Chromatography B**, v. 803, p. 237-241, 2004.

47. ROSERO, D. F. V. **Evaluation, production and quality of yuca fodder. *Manihot esculenta* Crantz, with periodic manual cutting**. 2002. 65 f. Dissertation (Degree in Agronomic Engineering) - National University of Colombia. Palmira. 2002.

48. MONTALDO, A. **Yuca or cassava**. San Jose: IICA, 1985.

49. ANTOV, M. G.; SCIBAN, M. B.; PETROVIC, N. J. Proteins from common bean (Phaseolus vulgaris) seed as a natural coagulant for potential application in water turbidity removal. **Bioresource Technology,** v. 101, p. 2167-2172, 2010.

50. ANTOV, M. G. et al. Investigation of isolation conditions and ion- exhange purification of protein coagulation components from common bean seed. **Acta Periodica Technologica**, v. 38, p. 3-10, 2007.

51. BENEVIDES, N. M. B.; LEITE, A. M.; FREITAS, A. L. P.. Hemagglutinating activity in the red alga Solieria filiformis. **Revista Brasileira de Fisiologia Vegetal**, v. 8, n. 2, p. 117-122, 1996.

52. MOREIRA, R. D. A.; PERRONE, J. C. Purification and partial characterization of a lectin from Phaseolus vulgaris. **Plant Physiology**, v. 59, p. 783-787, 1977.

53. BURGESS, R. M. **Protein purification.** In: Proteomics of the nervous system. NOTHWANG, S. E.(Ed.), Verlag GmbH & Co, Weinheim, 2008.

54. SCOPES, R. K. **Protein Purification. Principles and Practice**. 3. ed. New York: Springer-Verlag, 1993.

55. WANG, H. X.; NG, T. B. A novel lectin from Pseudostellaria heterphylla roots with sequence to Kunitz-type soybean trypsin inhibitor. **Life Sciences**, v. 69, p. 327-331, 2001.

56. REGO, E. J. L. et al. Lectins from seeds of Crotalaria pallida (smooth rattkebox). **Phytochemistry**, v. 60, p. 441-446, 2002.

57. KABIR, S. Jacalin: a jackfruit (Artocarpus heterophyllus) seed- derived lectin of versatile applications immunobiological research. **Journal of Immunological Methods**, v. 212, p. 193-211, 1998.

58. PEREIRA, C. A. **Hemagglutinin from cassava leaves (*Manihot esculenta* Crantz): partial purification and toxicity**. 2007. 43 f. Dissertation (Master's Degree in Agronomy) - Federal University of Lavras. Lavras, MG. 2007.

59. GASSENSCHMIDT, U. et al. Isolation and characterization of a flocculating protein from *Moringa*

oleifera Lam. **Biochimica et Biophysica Acta**, v. 1243, p. 577-481, 1995.

60. KATAYAN, S. et al. Preservation of coagulant efficiency of *Moringa Oleifera*, a natural coagulant. **Biotechnology & Bioprocess Engineering**, v. 11, p. 489-495, 2006.

61. NDABIGENGESERE, A.; NARASIAH, K. S.; TALBOT, B. G. Active agents and mechanism of coagulation of turbid waters using *Moringa oleifera*. **Water Research**, v. 29, n. 2, p. 703-710, 1995.

62. OKUDA, T. et al. Isolation and characterization of coagulant extracted from *Moringa Oleifera* seed by salt solution. **Water Research**, v. 35, n. 2, p. 405-410, 2001.

63. BELTRAN-HEREDIA, J. et al. Removal of Alizarin Violet 3R (anthraquinonic dye) from aqueous solutions by natural coagulants. **Journal of Hazardous Materials**, v. 170, p. 43-50, 2009.

64. NDABIGENGESERE, A.; NARASIAH, K. S. Quality of water treated using *Moringa Oleifera* seeds. **Water Research**, v. 32, p. 781, 1998.

65. MAKKAR, H. P. S. **Quantification in Tree and Shrub Foliage**. Dordrecht. Netherlands: Kluwer Academic Publishers, 2003.

66. HASLAM, E. Vegetable tannins - Lessons of a phytochemical lifetime. **Phytochemistry**, v. 68, p. 2713-2721, 2007.

67. CHAVAN, U. D.; SHAHIDI, F.; NACZK, M. Extraction of condensed tannins from beach pea (Lathyrus maritimus L.) as affected by different solvents. **Food Chemistry**, v. 75, p. 509-512, 2001.

68. HASLAM, E. **Plant Polyphenols:** Vegetable Tannins Revisited. Cambridge. England: Cambridge University Press, 1989.

69. DE BRUYNE, T. et al. Condensed vegetable tannins: Biodiversity in structure and biological activities. **Biochemical Systematics and Ecology**, v. 27, p. 445-459, 1999.

70. BATTESTIN, V.; MATSUDA, L. K.; MACEDO, G. A. Sources and applications of tannins and tannases in foods. **Alimentaçâo e Nutriçâo**, v. 15, n. 1, p. 63-72, 2004.

71. SCHOFIELD, P.; MBUGUA, D. M.; PELL, A. N. Analysis of condensed tannins: a review. **Animal Feed Science and Technology**, v. 91, p. 21-40, 2001.

72. MUELLER-HARVEY, I. Analysis of hydrolisable tannins. **Animal Feed Science and Technology**, v. 91, p. 3-20, 2001.

73. FRANCIS, G.; MAKKAR, H. P. S.; BECKER, K. Antinutrional factor present in plant-derived alternate fish feed ingredients and their effects in fish. **Aquaculture**, v. 199, p. 197-227, 2001.

74. HAGERMAN, A.; BUTLER, L. G. The specificity of proanthocyanidin-protein interactions. **The Journal of Biological Chemistry**, v. 256, p. 4494-4497, 1981.

75. REED, J. D. et al. Condensed tannins: A factor limiting the use of cassava forage. **Journal of the science food and agricuture**, v. 33, p. 213-220, 1982.

76. MARIE-MAGDELEINE, C. et al. In vitro effects of Cassava (Manihot esculenta) leaf extracts on four development stages of Haemonchus contortus. **Veterinary Parasitology**, v. 173, p. 85-92, 2010.

77. MONDOLOT, L. et al. Domestication and defense: Foliar tannins and C/N ratios in cassva and a close wild relative. **Acta Oecologica**, v. 34, p. 147-154, 2008.

78. AWOYINKA, A. F.; ABEGUNDE, V. O.; ADEWUSI, S. R. A. Nutrient content of young cassava leaves and assesment of their acceptance as an agreen vegetable in Nigeria. **Plant Foods for Human Nutrition**, v. 47, p. 21-28, 1995.

79. DUNG, N. G.; MUI, N. T.; LEDIN, I. Effect of replacing a commercial concentrate with cassava hay (*Manihot esculenta* Crantz) on the performance of growing goats. **Animal Feed Science and Technology**, v. 119, p. 271-281, 2005.

80. WANAPAT, M.; PURAMONGKON, T.; SIPHUAK, W. Feeding of cassava hay for lactating dairy cows. **Asian-Australian Journal of Animal Science**, v. 13, p. 478-482, 2000.

81. TEO, C. R. P. A. et al. Production and physico-chemical characterization of cassava leaf protein concentrate. **Revista Brasileira de Engenharia Agricola e Ambiental**, v. 14, n. 9, p. 993-999, 2010.

82. THANG, C. M.; LEDIN, L.; BERTILSSON, J. Effect of using cassava products to vary the level of energy and protein in the diet on growth and digestibility in cattle. **Livestock Science**, v. 128, p. 166-172, 2010.

83. HONG, N. T. T. et al. Effects of timing of initial cutting and subsequent cutting on yields and chemical composition of Cassava hay and its supplementation on lactating dairy cows. **Asian- Australian Journal of Animal Science**, v. 16, n. 12, p. 17631769, 2003.

84. HUE, K. T. et al. Effect of feeding fresh, wilted and sun-dried foliage from cassava (*Manihot esculenta* Crantz) on the performance of lambs and their intake of hydrogen cyanide. **Livestock Science**, v. 131, p. 155-161, 2010.

85. COLLINS, P. J.; YAZAKI, Y. **Tannin extraction and processing**. US Patent 5,417,888. 24 Jan. 1994. 23 May, 1995.

86. NACZK, M.; SHAHIDI, F. Extraction and analysis of phenolic in food. **Journal of Chromatography A**, v. 1054, p. 95-111, 2004.

87. HAGERMAN, A. E. Extraction of tannin from fresh and preserved leaves. **Journal of Chemical Ecology**, v. 14, n. 2, 1988.

88. COBZAC, S. et al. Tannin extraction efficiency, from Rubus Idaeus, Cydonia Oblonga and Rumex Acetosa, using different extraction techniques and spectrophotometric quantification. **Acta Universitais Cibiniensis Seria F Chemia**, p. 55-59, 2005.

89. FAO/IAEA. **Quantification of tannins in tree foliage**. IAEA. Vienna, p. 31. 2000.

90. GINER-CHAVEZ, B. I. et al. A method for isolating condensed tannins from crude plant extracts with trivalent ytterbium. **Journal of Science Food and Agriculture**, v. 74, p. 359-365, 1997.

91. ALONSO-AMELOT, M. E.; OLIVEROS, A.; ARELLANO, E. Exhaustive extraction of phenolics and tannins from some sun- exposed forbs and shrubs of the tropical Andes. **Ciência**, v. 13, n. 4, p. 429-439, 2005.

92. MAKKAR, H. P. S. et al. Gravimetric determination of tannins and their correlations with chemical and protein precipitation methods. **Journal of Science Food & Agriculture**, v. 61, p. 161-165, 1993.

93. OZACAR, M.; SENGIL, I. A. Evaluation of tannin biopolymer as an acoagulant aid for a coagulation of colloidal particles. **Colloids and Surfaces A: Physicochemical Engineering Aspects**, v. 229, p. 85-96, 2003.

94. YIN, C. Emerging usage of plant-based coagulants for water and wastewater treatment. **Process Biochemistry**, v. 45, p. 1437-1444, 2010.

95. QUAMME, J. E.; KEMP, A. H. **Stabble tannin based polymer compound**. US patent 4,558,080. 24 Jan. 1984. 10 Dec. 1985.

96. REED, P. E.; FINCK, M. R. **Modified tannin mannich polymers**.

US patent 5,659,002. 24 mar. 1995. Aug. 19, 1997.

97. LAMB, L. H.; DECUSATI, O. G. **Manufacturing process for quaternary ammonium tannate, avegetable coagulating and flocculating Agent**. US patent 6,478,986. Aug. 18, 2000. 12 Nov. 2002.

98. BELTRAN-HEREDIA, J.; SANCHEZ-MARTiN, J.; GÓMEZ- MUNOZ, M. C. New coagulant agents from tanninn extracts: Preliminary optimization studies. **Chemical Engineering Journal**, v. 162, p. 1019-1025, 2010.

99. RENAULT, F. et al. Chitosan for coagulation/flocculation processes - And eco-friendly approach. **European Polymer Journal**, v. 45, p. 1337-1348, 2009.

100. BELTRAN-HEREDIA, J.; SANCHEZ-MARTiN, J.; DAVILA- ACEDO, M. A. Optimization of the synthesis of a new coagulant from a tannin extract. **Journal of Hazardous Materials**, v. 186, p. 1704-1712, 2011.

101. ADACHI, Y. Dynamic aspects of coagulation and flocculation. **Advances in Colloid and Interface Science**, v. 56, p. 1-31, 1995.

102. BOLTO, B.; GREGORY, J. Organic polyelectroelites in water treatment. **Water Research**, v. 41, p. 2301-2324, 2007.

103. LEVISKÂ, T.; RAMO, J. Coagulation of wood extractives in chemical pulp bleaching filtrate by cationic polyelectrolytes. **Journal of Hazardous Materials**, v. 153, p. 525-531, 2088.

104. OLADOJA, N. A.; ALIU, Y. D. Snail shell as coagulant aid in the alum precipitation of malachite green from aqua system. **Journal of Hazardous Materials**, v. 164, p. 1496-1502, 2009.

105. ZHANG, J. et al. A preliminary study on cactus as coagulant in water treatment. **Process Biochemistry**, v. 41, p. 730-733, 2006.

106. SANCHEZ-MARTiN, J.; GONZALEZ-VELASCO, M.; BELTRAN-HEREDIA, J. Surface water

treatment with tannin- based coagulants from Quebrancho (Schinopsis balansae). **Chemical Engineering Journal**, 2010. In press.

107. SZYGULA, A. et al. The removal of sulphonated azo-dyes by coagulation with chitosan. **Colloids and Surfaces A: Physicochemical and engineering aspects**, v. 330, p. 219-226, 2008.

108. COAGULATION/FLOCCULATION. Available at:

<http://www.universoambiental.com.br/Arquivos/Agua/Processos Wastewater Treatment Chemicals07.pdf>. Accessed on: August 24, 2011.

109. SEMERJIAN, L.; AYOUB, G. M. High-Ph-magnesium coagulation-flocculattion in wastewater treatment. **Advances in Environmental Research**, v. 7, p. 389-403, 2003.

110. ABDELAAL, A. M. **Using a natural coagulant for treating wastewater**. In: International Water Technology Conference. IWTC, 8. 2004, Alexandria. p. 781-792.

111. ALI, G.; EL-TAWEEL, G.; ALI, M. A. The cytoxixity and antimicrobial efficiency of *Moringa oleifera* seeds extracts.

International Journal of Environmental Study, v. 61, n. 6, p. 699-708, 2004.

112. JAYARAM, K. et al. Biosorption of lead from aqueous solution by seed powder of Strychnos potatorum L. **Colloids and Surfaces B: Biointerfaces**, v. 71, p. 248-254, 2009.

113. SANGHI, R.; BHATTACHARYA, B.; SINGH, V. use of Cassia javahikai seed gum and gum-g-polyacrylamide as coagulant aid for the decolorization of textile dye solutions. **Bioresource Technology**, v. 97, p. 1259-1264, 2006.

114. ANASTASAKIS, K.; KALDERIS, D.; DIAMADOPOULOS, E. Flocculation behavior of mallow and okra mucilage in treating wastewater. **Desalination**, v. 249, p. 786-791, 2009.

115. DI BERNARDO, A.; DI BERNARDO, L. **Use of cationic cassava starch as a flocculation aid**. In: Inter-American Congress of Sanitary and Environmental Engineering, 15. 2000, Rio de Janeiro: ABES - Brazilian Association of Sanitary and Environmental Engineering, 2000. p. 1-12.

116. CRUZ, J. G. H. et al. **Application of a vegetable coagulant based on tannin in the treatment by coagulation/flocculation and adsorption/coagulation/flocculation of effluent from an industrial laundry**. In: Congresso Brasileiro de Engenharia Sanitaria e Ambiental, 23. 2005, Campo Grande: Associaçao Brasileira de Engenharia Sanitaria e Ambiental, 2005. p. 1-12.

117. PRITCHARD, M. et al. Potential of using plant extracts for purification of shallow well water in Malawi. **Physics and Chemistry of the Earth**, v. 34, p. 799-805, 2009.

118. PATEL, H.; VASHI, R. T. Removal of congo red dye from its aqueous solution using natural coagulants. **Journal of Saudi Chemical Society**, p. In press., 2011.

119. RAGHUWANSHI, P. K. et al. Improving filtrate quality using agrobased materials as coagulant aids.

Water Quality Research Journal Canada, v. 37, n. 4, p. 745-756, 2002.

120. SANTANA, C. R. et al. Evaluation of the process of coagulation/flocculation of produced water using *Moringa oleifera* Lam. as natural coagulant. **Brasilian Journal of Petroleum and Gas**, v. 4, n. 3, p. 111-117, 2010.

121. BONGIOVANI, M. C. et al. The benefits of using natural coagulants to obtain drinking water. **Acta Scientarium**, v. 32, n. 2, p. 167-170, 2010.

122. GHEBREMICHAEL, K. A. et al. A simple purification and activity assay of the coagulant protein from *Moringa oleifera* seed. **Water Research**, v. 39, n. 11, p. 2338-2344, 2005.

123. GHEBREMICHAEL, K. A.; GUNARATNA, K. R.; DALHAMMAR, G. Single-step ion exchange purification of the coagulant protein from *Moringa oleifera* seed. **Applied Microbiology and Biotechnology**, v. 70, p. 526-532, 2006.

124. KATAYON, S. et al. Effects of storage conditions of *Moringa oleifera* seeds on its performance in coagulation. **Bioresource Technology**, v. 97, p. 1455-1460, 2006.

125. SOCIETY OF DYERS AND COLORISTS. Colorants. In: SHORE, J. **Colorants and auxiliaries:** organic chemistry and application properties. Yorkshire, England: Society of Dyers and Colourists, v. 1, 2002. p. 372.

126. ABOULHASSAN, M. A. et al. Improvement of paint effluents coagulation using natural and synthetic coagulant aids. **Journal of Hazardous Materials B**, v. 138, p. 40-45, 2006.

127. MISHRA, A.; BAJPAI, M. The flocculation performance of Tamarindus mucilage in relation to removal of vat and direct dyes. **Bioresource Technology**, v. 97, p. 1055-1059, 2006.

128. ZAHRIM, A. Y.; TIZAOUI, C.; HILAL, N. Coagulation with polymers for nanofiltration pre-treatment of highly concentrated dyes: A review. **Desalination**, v. 266, p. 1-16, 2011.

129. BELTRAME, L. T. C. **Microemulsion systems applied to the removal of color from textile effluents**. 2006. 185 f. Thesis (Doctorate in Chemical Engineering) - Federal University of Rio Grande do Norte. Natal. 2006.

130. RAFATULLAH, M. et al. Adsorption of methylene blue on lowcost adsorbents: A review. **Journal of Hazardous Materials**, v. 177, p. 70-80, 2010.

131. BIBI, I.; BHATTI, H. N.; ASGHER, M. Comparative study of natural and synthetic phenolic compound as efficient laccase mediator for the transformation of cationic dye. **Biochemical Engineering Journal**, v. 56, p. 225-231, 2011.

132. DOGAN, M. et al. Adsorption kinetics of maxilon blue GRL onto sepiolite from aqueous solution. **Chemical Engineering Journal**, v. 124, p. 89-101, 2006.

133. DRAGONTECH & WALLERT AND PROVOST LAB. Ammonium Sulfate PPT. Protocol, November 5, 2009. Available at: <http://www.dragontech.com>. Accessed on: August 8, 2011.

134. ASSOCIATION OF OFFICIAL ANALYTICAL CHEMISTS - AOAC. **Official methods of analysis of AOAC International.** 16. ed. Arlington: Association of Analytical Communities, 1995.

135. BRADFORD, M. M. A rapid and sensitive method for the quantitation of microgram quantities of protein utilizing the principle of protein-dye binding. **Analytical Biochemistry**, v. 72, p. 248-254, 1976.

136. BARMAN, K. et al. Tannins estimation, 2004. Available at:< http://pt.scribd.com/doc/43638556/6506867-Tannin-Assay>. Accessed on: September 6, 2001.

137. PORTER, L. J.; HRSTICH, L. N.; CHAN, B. J. The conversion of proanthocyandins oridelphinidins to cyanidin and delphinidin. **Phytochemistry**, v. 223-330, p. 25.

138. CEROVIc, L. S. et al. Point of zero charge of different carbides. **Colloids and Surfaces A: Physicochemical Engineering Aspects**, v. 297, p. 1-6, 2007.

139. AINOUZ, I. L. et al. Agglutination of enzyme treated erythrocites by Brazilian marine Algae. **Botanica Marina**, v. 35, p. 475-479, 1992.

140. AMERICAN PUBLIC HEALTH ASSOCIATION. **Standard Methods for the Examination of Water and wastewater**. 20. ed. Washington D.C.: American Public Health Association, 1998.

141. PÉREZ, O. P.; LAZO, F. J. *Artemia* testing: a useful working tool for ecotoxicologists and chemists of natural products. **Revista de Protección Vegetal**, v. 22, n. 1, p. 34-43, 2010.

142. TEÓFILO, R. F.; FERREIRA, M. M. C. Quimiometria II: Spreadsheets for experimental design calculations, a tutorial. **Quimica Nova**, v. 29, n. 2, p. 338-350, 2006.

143. COATES, J. **Interpretation of infrared spectra, a practical approach.** In: Encyclopedia of Analytical Chemistry MEYERS, R. A.; (Ed.), Wyley 6 Sons, 2000.

144. KHAN, T. A.; QAYYUM, H.; AABGEENA, N. A novel and inexpensive procedure for the purification of concavalin A from jack bean (*Canavalia ensiformis*) extract. **European Journal of Applied Sciences 2**, , v. 2, p. 70-76, 2010.

145. PADMAJA, G. Evaluation of techniques to reduce assayable tannin and cyanide in cassava leaves. **Journal of Agricultural and Food Chemistry**, v. 37, p. 712-716, 1989.

146. CAMARA, F. S.; MADRUGA, M. S. Cyanic acid, phytic acid, total tannin and aflatoxin contents of a Brazilian (Natal) multimixture preparation. **Revista de Nutriçâo de Campinas**, v. 14, n. 1, p. 33-36, 2001.

147. CHAKRABARTI, A.; PODDER, S. Complex carbohydrate-lectin interaction at the interface: a model for cellular adhesion. I. Effect of vesicle size on the kinetics of aggregation between a fatty acid conjugate of lectin and a liposomal asialoganglioside. **Biochimica et Biophysica Acta**, v. 1024, n. 1, p. 103-110, 1990.

148. SOBRAL, P.; PALAZOLO, G. G.; WAGNER, J. R. Thermal behavior of soy protein fractions depending on their preparation methods, individual interactions, and storage conditions. **Journal of Agricultural and Food Chemistry**, v. 58, n. 18, p. 1009210100, 2010.

149. YAN, M. et al. Mechanism of natural organic matter removal by polyaluminum chloride: effect of coagulant particle size and hydrolysis kinetics. **Water Research**, v. 42, p. 3361-3370, 2008.

150. KIM, T. et al. Decolorization of disperse and reactive dye solutions using ferric chloride. **Desalination**, v. 161, p. 49-58, 2004.

151. MALL, I. D. et al. Adsorptive removal of malachite green dye from aqueous solution by bagasse fly ash and activated carbonkinetic study and equilibrium isotherm analyses. **Colloids Surface A: Physicochemical Engineering Aspects**, v. 264, p. 17-28, 2005.

ANNEX 1.

Bovine serine albumin (BSA) standard curve

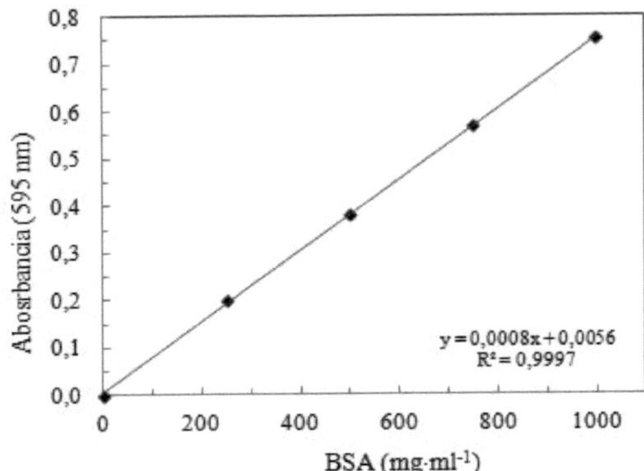

Tannic acid standard curve.

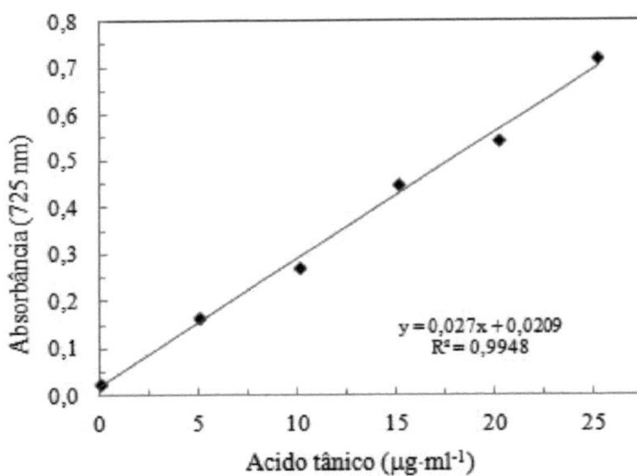

AM standard curve.

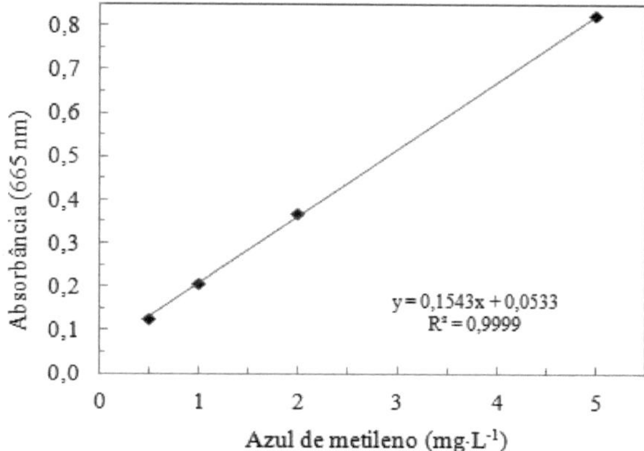

Standard AMG curve.

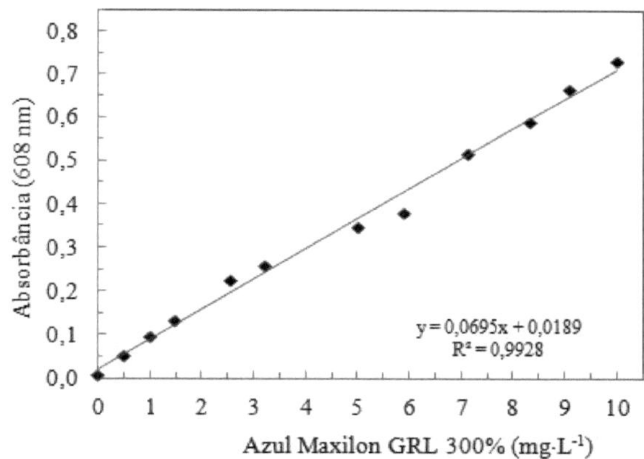

Standard MV curve.

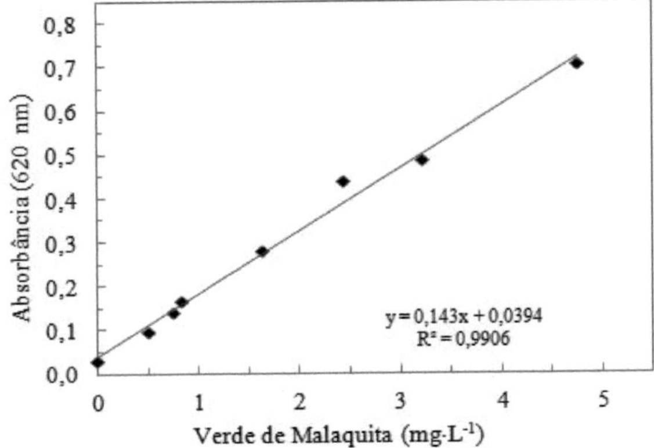

Potassium hydrogenphthalate standard curve

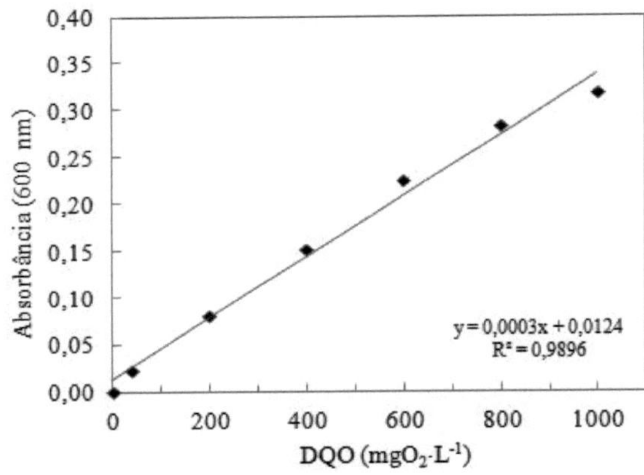

Zero charge point (pHCZ) determination curve for extract E5.

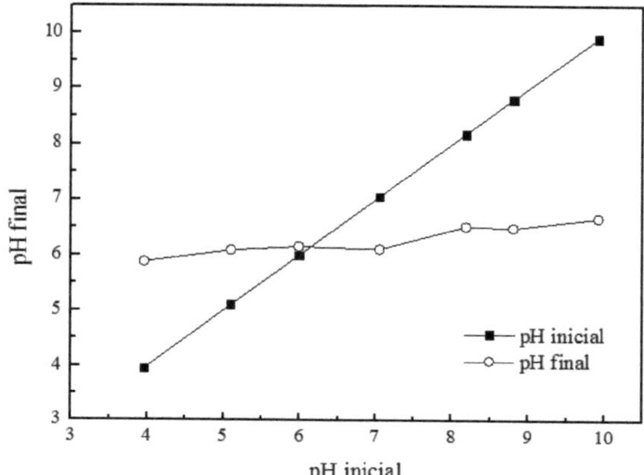

Printed by Books on Demand GmbH, Norderstedt / Germany